STAMPEDE

Saints, Successes, Suckers & Scoundrels of the Yukon Gold Rush

SAMSON PRESS

Chugiak, Alaska

ISBN-978-0-9721166-5-7

Jordan, Lee B.
1930 -

STAMPEDE! Saints, Successes, Suckers &
Scoundrels of the Yukon Gold Rush
Includes Index

Other Books by Lee Jordan

Starlight Memories, a collection of his favorite columns from the *Chugiak-Eagle River (Alaska) Star.*

Reflections of a Reluctant Alaskan, Twelve times he responded to the Army's question as to his choice of overseas assignment saying, "anyplace but Alaska." On January 4, 1949, he stepped off a troopship in Whittier at 20 below zero. He is still there, reflecting on a joyful life.

Cover photo:

Stampeders at The Scales in September, 1898, climbing the "Golden Staircase" of the Chilkoot Pass rising from Dyea over the mountains to the waterways that lead to the Yukon town of Dawson. Photograph by E.A. Hegg
University of Washington Libraries, Special Collections, HEG 408

For

all those who dared

to chase the rainbow

in search of their dreams

and especially

to those who remained to make

the North truly the Great Land that it is.

Aerial view of Nome in 1900 as the Anvil Creek stampede was just getting underway. Tents sprang up on the beach to house newcomers as the population swelled.

Photographer unknown

Illustrations

Note: Photos not otherwise credited are from royalty-free sites in the public domain.

Table of Contents

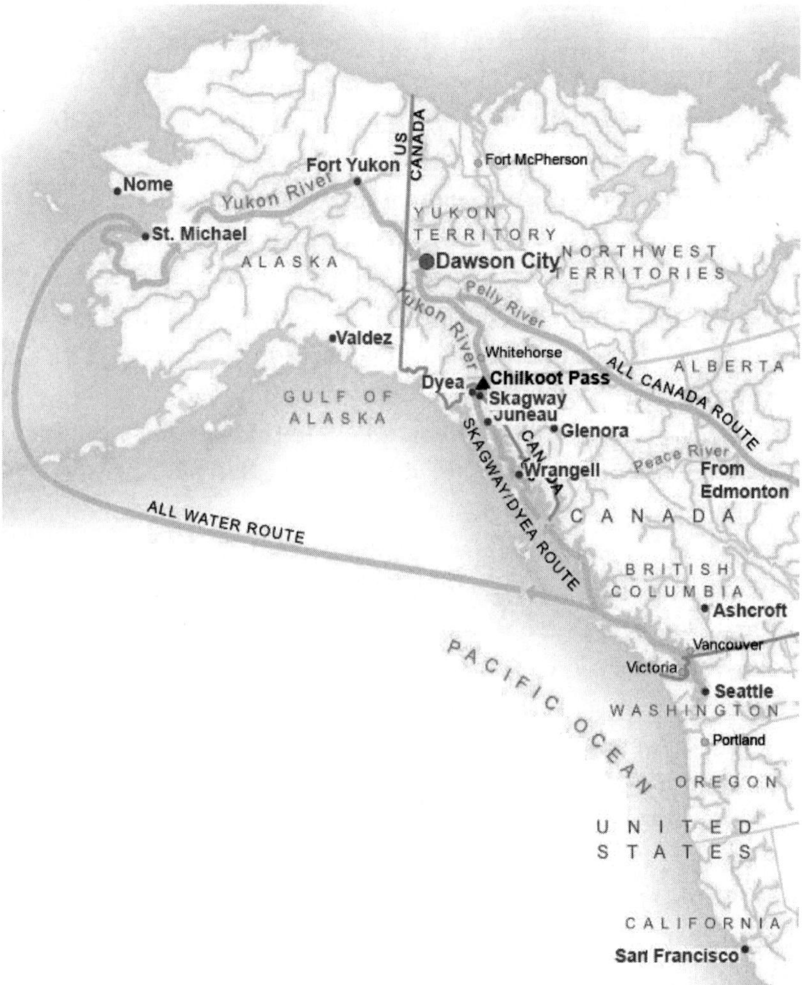
US
CANADA
Fort McPherson
Fort Yukon
Nome
Yukon River
St. Michael
YUKON TERRITORY
ALASKA
Dawson City
NORTHWEST TERRITORIES
Pelly River
Yukon River
Valdez
Whitehorse
ALBERTA
ALL CANADA ROUTE
Dyea
Chilkoot Pass
Skagway
GULF OF ALASKA
Juneau
Glenora
SKAGWAY/DYEA ROUTE
Peace River
Wrangell
From Edmonton
CANADA
ALL WATER ROUTE
BRITISH COLUMBIA
Ashcroft
Vancouver
Victoria
PACIFIC OCEAN
Seattle
WASHINGTON
Portland
OREGON
UNITED STATES
CALIFORNIA
San Francisco

Acknowledgements

The author is indebted to a great many people who have contributed to this work. First, the Internet is a resource where abounding information on the Gold Rush can be found. In addition, hundreds of books on the subject have been published, scores of which are in his own collection of Alaskana. Foremost among them is Pierre Berton's "Klondike Fever." The son of a Klondike stampeder, Berton was born in Whitehorse, Yukon Territory, and prior to his death in 2004 at age 84, had written fifty books primarily on Canadian history. Next in importance as a reference is Russell A. Bankson's "The Klondike Nugget," the story of Dawson's first newspaper and its publisher, Eugene Allen.

An inspiration lies in the life of the Rev. Dr. Sheldon Jackson, a Presbyterian minister who was appointed as General Agent for Education for Alaska and who devoted his life to improving the lot of Alaska Natives—a people he first thought of as "heathens" but who he came to respect. He put into action a project suggested by the inimitable Captain "Hell-roaring" Mike Healy, commander of the Revenue Cutter *Bear*, to introduce reindeer as an industry as well as to supplement the diet of the indigenous people of the northland.

A big "thank you" is due to Michael Hawfield, associate professor emeritus of history at the Kachemak Bay campus of the University of Alaska Anchorage for writing the foreword to this work.

A personal interest in placer mining came from Walter A. Erickson, the father of the author's bride and the son of John T. Erickson, a stampeder who first went to Nome to mine for gold, then went back with a herd of horses and mules from which he profited even more. Walter Erickson invented an undercarriage for a washing plant and did exploration work on claims at Nome during the 1950s. In 1939 he had flipped a coin to determine whether he followed his father's footsteps to Alaska or chased the rainbow in Australia, the result being fortunate for the author.

A note of appreciation goes to the staff of the Carrie M. McLain Memorial Museum at Nome, the Consortium Library on the campus of the University of Alaska Anchorage, and the Digital Archivists at the University of Washington who provided many of the photographs used to illustrate the lives and times of the stampeders whose stories are told here. Also of help was Adria Lund of the Glenbow Museum, Calgary, Alberta, Canada.

The author would be remiss if he did not give credit to the work of the Alaska Mining Hall of Fame, whose inductees include many of the successes who are among the characters described herein.

Of course, the author's undying gratitude goes to Barbara, who became his bride 65 years ago, for her patience and understanding during the years leading up to the publication of this work. She has accompanied him on many visits to the trails the stampeders trod so long ago.

Foreword

No event in Alaska has produced more fascinating, heroic, bogus, heartbreaking, fantastic, villainous, distressing, amazing, tragic, fruitless, or wondrous stories than the great northern Gold Rush at the turn of the Twentieth Century. Of course, there are important historical studies examining such topics as the role of the Alaska Commercial Company, the canneries, or the demographic characteristics of the thousands who rushed north pursuing dreams of finding gold, but the tales surrounding the individuals and their extraordinary exploits reveal a deeper, personal connection with the phenomenon of the Stampede of '98. The Stampede and the Alaska experience generated a rich array of poems, short stories, novels, and films. What impressed these authors about all that they experienced in the Gold Rush is also what has drawn newspaperman Lee Jordan to the characters and events of the Stampede throughout the region of the Yukon River.

Following in the journalist traditions of Pierre Burton and his classic work, *Klondike Fever,* Lee Jordan is a master story-teller. It is what he has done for more than six decades, including as a columnist and editor of the *Chugiak-Eagle River Star.* Many readers will recall with delight his popular *Reflections of a Reluctant Alaskan,* which revels in the allure of Alaska, despite the hardships and absurdities of actually living there. Jordan embraces the Gold Rush with the same enjoyment and sense of humor in the wonderful stories that define the era. In this book, he gives us sixty-five lively historical vignettes touching a rich variety of topics, ranging from famous and not-so-famous individuals to important events and eccentric episodes.

This is not a "linear" history of the Gold Rush, seeking to describe from beginning to end what happened, but a marvelous collage of stories that bring this grand event to life. Reflecting several years of research, this book is in some ways a "file cabinet between two covers," for the reader is pleased at every "tab" in the file to find another topic of interest. When I went to the first section, entitled "Klondike Strike," I found accounts of some of the core stories – the discoveries on the Klondike River by the Carmack family, the exciting news as it reached Outside, and the hubbub and shenanigans of the boom towns – but I soon also discovered a host of less common tales. I was delighted to find, for example, the stories of the lonely death of the renowned prostitute "Diamond Lil," the stranded stampeders of "Suckerville," the tough business woman named Belinda Mulrooney who grounded her fortune in hot-water bottles, the beloved hotelier "Ma" Pullen of Skagway, and the fascinating tale of Ed and Velma Lung who outlasted the all-too-common legal challenges to their claim and struck it rich.

True to his background in journalism, one of the important contributions to our understanding of the Gold Rush era are Jordan's stories of newspapers, newsmen and photographers. He opens his book with a story about the news of the great Klondike strike and how the news accounts themselves fueled the frenzy. Notably, he recounts the story of Eugene Allen, the editor and publisher of the *Klondike Nugget*, which is a mine of stories and information about life on the trail and in the gold camps. The Stampeders called themselves “Argonauts” in comparison to the ancients who followed Jason to find the “golden fleece.” Allen was like all the other Argonauts, broke and full of ambition and hope. Undeterred, he and his partners lugged the huge bulk of printing equipment over the Chilkoot and White passes and down the Yukon River to Dawson and there began to capture the stories as they unfolded. With these stories he and columnist “Stroller” White established the most important record of the Stampede in Dawson, but also assumed the role of a classic muckraking journalist exposing corruption in local government and officialdom at all levels.

One of the enjoyable aspects of this book is how the reader comes across many familiar topics but finds a fresh telling. For instance, the story of “Klondike Kate” Rockwell provides a classic tale of the young, adventurous woman who was drawn to the gold fields not to prospect in the creeks, but to experience the excitement of the Stampede and to win her fortune in the dance halls. Kate’s loves and heartbreaks became the stuff of legend on the Yukon as she danced and flirted at “Arizona Charlie’s” Grand Opera House. When she was honored at a reunion of Argonauts in the 1930s, Jordan tells us, “one thousand of the sourdoughs rose to their feet and applauded loudly and long.” Her fellow entertainer at the Grand Opera House, “Diamond Tooth” Gertie, we learn later, also gained fame and wealth, but we also meet in other stories the extraordinary women who baked or laundered their way to the camps.

Jordan also explores the frenzied rushes down the Yukon, to Forty Mile, and up on the Seward Peninsula, to Nome, which was plagued by the monumental scams that defined the activities of Alexander McKenzie and crooked Judge Arthur Noyes. Here, Jordan's own deep familiarity – and family connections – with prospecting and mining laws help sort out the intricacies of staking claims and claim-jumping, and makes clear the importance of Judge James Wickersham who came to the boom town to help the swindled miners find justice and established the foundations for his own storied political career as the foremost Alaska booster. Focusing on the western part of Alaska in that era also gives Jordan the opportunity to examine how famous wild-west figures like Wyatt Earp (who came to be a hotelier) and John Clum (among many other things in his life, editor of the *Tombstone Epitaph*) were drawn to the gold camps, and how the novelist and short-story writer Rex Beach was inspired to write some of his most popular stories.

In the final sections of the book the reader is reminded that the Gold Rush era involved much more than the frenzied "search for color." We meet missionaries like Sheldon Jackson and his scheme to introduce reindeer to Alaska as a result of his perception that Alaska Natives needed a more abundant source of food. Native people of course continued to harvest their traditional foods, but the reindeer also proved an unexpected boon to stranded whalers and hungry miners. We are also introduced to "Hell-Roaring" Mike Healy, skipper of the U.S. Revenue Cutter *Bear,* who was chief law-enforcement officer of the territory, and "Cassiar Angel" Nellie Cashman, who defied the Canadian military to save stranded Stampeders in the deep snows of the Cassiar Mountains.

Among the enduring images of Alaska are the sled dogs. Mushing was important to both Alaska Natives and to the more intrepid Gold-Rush travelers along the Yukon, not the least of whom were the federal circuit judge James Wickersham and the hearty mail carriers. It's not surprising that the four-legged champions of the Yukon also became the centerpiece of the uniquely Alaskan tradition of dog-racing and the central character of one of Jack London's most popular Gold-Rush novels.

Thanks to the Gold Rush along the length of the great Yukon River, Alaska gained global attention. The national government, too, gradually began to embrace its unique northern territory, but not before the spirit of the Gold Rush – with its excitement, promise, and the indomitable spirit of the thousands of participants, as revealed in their stories – had forged a lasting image of a great land, both challenging and beautiful.

Lee Jordan's wonderful collection of stories about the men and women of the amazing Gold Rush era along the Yukon River are short, lively, and well informed. Readers will find them highly readable and guaranteed to keep them turning the pages to find the next fascinating tale. The only disappointment will be discovering there are no more when the last page is turned.

Michael Hawfield
Associate Professor Emeritus of History
Kenai Peninsula College
University of Alaska Anchorage

This scene faced Stampeders approaching the new town of Dawson in 1898. The town was laid out by trader Joseph Ladue soon after the discovery of Gold on Rabbit Creek two years before. Photo courtesy Library of Congress, 3b12760r

Part I

The Klondike Stampede

Dyea waterfront was busy in 1898 with passengers and goods unloaded from ships anchored off-shore. Dyea was the beginning of the Chilkoot Trail, gateway to the Klondike.

Eric A. Hegg Photo, courtesy National Park Service

Once over the passes, stampeders had to navigate streams to reach the Klondike. Here three men on raft make their way through Miles Canyon.

University of Washington, Special Collections, UW 17194

Introduction

They were a disparate group: A priest who wanted only to serve people in need, a fruit farmer unable to afford to keep his orchards producing, an adventure-seeking young man hood-winked into joining the rush to the Klondike, and a nuevo-riche n'er-do-well with a penchant for young girls. From different backgrounds, from different nations, and with different motives, thousands of people found themselves thrown together in the frontier town of Dawson in the northwest corner of North America. It was 1898, the year when an estimated one hundred thousand people rushed toward the spot where gold had been found in great quantities. A few actually struck it rich. Thousands were disappointed and forced either to work for wages or turn back. Hundreds lost their lives in the desperate effort to realize their dreams.

Still recovering from the depression that ravaged the nation in 1893, their emotions had been stirred months earlier by headlines that read, "A TON OF GOLD!" The northwestern edge of the continent had garnered little attention since Alaska was purchased twenty-one years earlier. "Seward's Folly" was still considered to be nothing more than that. Other than adventurers set on exploring this new land, and missionaries seeking to spread the Gospel to the "ignorant heathens" there, the far north held little interest.

That changed in July of 1897 when the steamer *SS Excelsior* arrived in San Francisco, California, and the *SS*

Portland docked in Seattle, Washington. Both ships carried miners who had become wealthy by working claims in the Klondike region. Gold had been discovered there a year earlier, a fact until now known only to those who were already prospecting along the Yukon. The bags of gold carried by those miners caught the attention of newspapermen who were anxious to spread the word.

Only a few of those who joined the stampede hurrying to an unknown future in an unknown land had an inkling of what to expect. Their focus was on getting to this suddenly fabled place where glittering gold nuggets were lying on the ground, just waiting to be picked up.

Little did they know the challenges to be faced and overcome, merely to get near the creeks that harbored all this wealth. There were mountains to be crossed on foot, lakes to be traversed, rushing rivers with raging rapids waiting to destroy all who dared to venture along their route. Yet to be realized was that by the time they finally reached their destination, all the promising streams had been staked for their entire limit.

Together, those who did not turn back found ways to profit from their efforts. They forged a fellowship from just surviving the journey, a bond that lasted. Their character had been tested. It was they who created a place that was to live and thrive for centuries to follow.

The great Yukon, with its headwaters in Canada's Yukon Territory and wandering for two thousand miles across Alaska to empty into the Bering Sea, was to be made famous by a collection of saints, successes, suckers and scoundrels. We are fortunate to gain some insight into the lives of the people who were a part of this great adventure.

News of Klondike discovery gains notice worldwide

News of the Klondike gold discovery spread quickly around the world. On July 17, 1897, headlines screaming "A TON OF GOLD!!!" stirred the imagination of people everywhere and spurred the adventurous to action. It is estimated that one hundred thousand people quickly headed for the Klondike. Forty thousand of them made it to the new settlement of Dawson and that camp's environs. Thousands of the original stampeders either turned back or stopped to explore other creeks. Scores of thousands more were to rush northward when word came that three "lucky Swedes" had made another fabulous discovery at the other end of the mighty Yukon.

Conditions were right to spawn the stampede that followed. The effects of an economic depression that began in 1893 were still being felt. News from Chicago of its big Columbian Exposition, combined with clouds of war stemming from unrest in Cuba and the Philippines, had people thinking about things beyond their own familiar spheres of activity. Westward growth related to the California gold rush of '49 came to mind for entrepreneurs who saw opportunity arising from this new possibility.

News stories about the big Klondike strike in many cases were exaggerated in follow-up articles written by reporters who had been scooped by the initial coverage. Editors fashioned headlines designed to grab attention to their newspapers, vying with many competitors for sales. Cries on the streets of "EXTRA! EXTRA!" made the big event a universal topic of conversation.

The discovery of gold in the northern climes was nothing new. Prospectors began exploring the mysterious far north even before Alaska was purchased from Russia in 1867. Streams in remote areas of both Alaska and Canada produced small amounts of the yellow metal, encouraging further exploration. By 1886, camps had been established at Fortymile, near the Yukon River border between Alaska and Canada and several other areas in both nations. The first major gold camp on the Alaska side was at Circle City after a sizeable find on Birch Creek in 1893.

A New Yorker seeking the fortune he hoped would gain his sweetheart's hand had been traversing the creeks feeding the mighty Yukon River for four years. Joseph Ladue had pursued many professions before establishing an Alaska Commercial Company trading post at Fortymile. There, in addition to selling groceries and tools, satisfying his customers' thirst and bartering for furs, he also whip-sawed lumber from the scraggly spruce that grew in the area, selling planks to miners for use in building boats, cabins and sluice boxes. Little did he suspect that he would play a huge part in the future of the Northland.

One summer day in 1896, a group of miners were in Ladue's saloon, enjoying a meal and washing it down with spirits from the well-stocked bar, when a tall man strode in and sat down. He was recognized as the one known around the area upriver as "McCormick," an American who lived with the Indians and was generally considered to be someone whose veracity was not always reliable. Drawn into conversation, the newcomer disclosed that he had just filed discovery claims on Rabbit Creek, which he now called Bonanza, near the confluence of the Klondike and Yukon rivers some fifty miles upstream from where they sat. Having heard a little about the man's history, the listeners cautiously questioned his story. To their astonishment, he pulled out a shotgun shell filled with coarse gold that he had brought along to prove to Canadian authorities that he had found gold in sufficient quantity to justify filing a claim.

The man known as "McCormick" was actually George Washington Carmack, who had come north some years earlier from California. He became friendly with the Natives around Lake Tagish and had taken as his common-law wife a young widow who was the daughter of a tribal chief of the Raven clan. An intelligent and attractive woman, Shaaw Tláa adopted the name Kate Carmack and accompanied him as he roamed the creeks whose waters dumped into the Yukon. As related to the men at the bar, "Skookum" Jim, Kate's brother, and her nephew, "Indian Charlie," were traveling with Carmack when he met Robert Henderson on Indian Creek, a tributary of the Yukon. In the custom of prospectors of that day, Henderson told them of his find and recommended that they check a nearby creek where good color had been found. Moving a little farther upstream to Rabbit Creek, they scratched material from the cold water and swished it in a pan. Carmack was amazed to see a quarter of an ounce of gold glistening in the sands of the pan held in his hand. At four dollars, that discovery was phenomenal when considering ten cents a pan at that time indicated a highly favorable prospect.

Carmack immediately scraped the bark from an alder

alongside the creek and on the bared trunk scribbled out a notice stating his discovery claim. His in-laws each measured out and staked adjoining claims.

That was the story told to listeners at Fortymile. Later, however, several other versions began to circulate. Carmack was known to be somewhat lazy and rather than expend effort in digging for gold was content to have his wife and her relatives do that while he fished and contemplated the scenery. Kate, her brother and nephew were later variously credited with being the actual discoverers, with Carmack taking the credit. The reasons for the controversy are debatable. Canadian historians later listed fellow countryman Henderson as an official co-discoverer while boosters of the First Nation movement pressed for the inclusion of Shaaw Tláa along with them as a discoverer. With national and cultural angles as well as morality and justice involved, the "discoverer" debate raged well beyond the Yukon and across borders.

Carmack spent some time checking other creeks for signs of the yellow metal before heading for Fortymile to file his claim. Along the way, he showed off the vial of gold and told others he met of the discovery—a courtesy generally afforded to those who shared the privations experienced by early explorers in the north. One with whom he did not share the news, however, was Henderson, the man who had pointed the Carmack party to the spot which was soon to become known as the richest creek in the world.

The consensus of many of those who recorded the history of the Yukon is that Carmack was angered by Henderson's reaction when the visitors expressed interest in staking claims alongside his. Despite the man's location on Indian Creek, Henderson had no desire to have neighbors who were members of the First Nation. He reportedly used rather offensive language in making that opinion known. Henderson, who hailed from Nova Scotia, had gone to sea at age 14, two decades before he headed for the Yukon. He felt he was always plagued by bad luck and was embittered at being ignored for his role while the ingrate Carmack was hailed as the Klondike discoverer.

Whatever the real story of the discovery might be, what followed was the biggest gold rush the world had ever seen. It gained a place in history that rivals all others for adventure, glory, wealth, romance, fame, notoriety, greed, corruption, lawlessness, heroics, and, yes, continued controversy.

The Great Stampede of '98 began to die off after six or seven years, but the northern part of the continent continued to produce gold, and does so to this day. Huge machinery replaced hand shovels, science replaced on-site experiments, consolidation by large corporations replaced individual ownership of claims, and huge dredges replaced sluice boxes and rockers.

Fortunes were made by the hardy pioneers who were lucky enough to strike pay dirt. Gold back then was worth sixteen to seventeen dollars an ounce, depending on its quality. Although described as yellow in color, the hue varies depending upon influence of other minerals in the creek from which it is taken. Experienced miners can actually identify with a fair degree of certainty the locality from which a gold specimen is taken.

In 2011, gold's price rose to almost nineteen hundred dollars per ounce. It has declined since, but remains high enough to generate considerable interest. In addition to increased price, conditions facing miners certainly have improved. Refinements in equipment and methods have turned prospecting into recreational trips. Remote creeks can be reached within not months but hours from anywhere on earth. Working in ice-cold water can be done in relative comfort. Living arrangements can be made far more comfortable.

Anyone today who thinks about jumping on an airplane with a gold pan and a sack lunch, though, needs to consider other things that now are different from those of more than a century ago.

Most undeveloped land presently is either under government control or in private ownership. Research is necessary to determine what areas might be open for prospecting. Areas that have been set aside for recreational mining use are few and in great demand. The Nome beach that was so productive and open to everyone in 1898, for example, can still be mined for short periods—-but only by permit. In 2013, the City of Nome found it necessary to limit parking near the beach due to the high demand that brought overcrowding.

Cities have grown up around the mining camps of old. Those cities now are inhabited by residents who have concerns that were not of importance when gold fever gripped people who came from afar to seek their fortunes. Turning forests into dust bowls is no longer forgivable as it was when timber was needed to build boats, or when the trees got in the way of getting to the gold. People want streams to be clear; they don't want them polluted by chemicals or the water to become so fouled that fish cannot survive. To insure that their concerns are addressed, the public has persuaded politicians to write clean-water laws—and bureaucrats have promulgated regulations that often go well beyond the wording of the laws.

Those caveats are listed for the benefit of readers who

might be as tempted to head to the backcountry as were those souls in the days of yore. Only the hardiest got through. Only the luckiest of them came back with anything of value other than memories—and exciting stories to be told. For those of us who are interested in those stories, a retelling gives an opportunity for a glimpse into what they experienced.

Presented here is a look at some of those pioneers. Only a handful of them were experienced miners. The rest included men and women who saw opportunities worth risking one's life and life savings; fortune-hunters who were willing to toil to earn those fortunes, or perhaps to wrench fortunes from others by fair means or foul; entertainers who saw a venue for even the less talented in an area where men had few choices but time on their hands—and gold to spend; lawyers who realized that disputes would be plentiful in a new frontier where even the few laws that were on the books were themselves in dispute; and chroniclers of the era.

It is the latter in that cast of characters who today let us peek into the lives of the people who joined the Great Stampede. Photographers lugged heavy cameras along the trail, setting them up to take the best advantage of natural light, coating glass plates with emulsion they mixed in the dark and carefully protected, then getting their subjects to pose without moving while the exposure was timed. Writers whose instinct told them there would be great stories to be recorded, today as well as then, proved to be valuable sources for those who could not themselves take part in the Grand Adventure.

Contemporary writers who detailed the exploits of those who ventured north in search of gold almost universally referred to them as "argonauts." The term was resurrected from the tales of the Greek hero, Jason, who went in search of the golden fleece of mythology. His group sailed on the vessel *Argo,* hence the name. As this work is intended to be more factual than mythical, the more fanciful term will be rarely used in favor of adventurer, stampeder, miner, prospector, or other such term.

This effort will look at a sampling of the people who took part in this important part of history. The mass that rushed to the north made Alaska and the Yukon what they have become. Their spirit lives on in lore and in fact. Their like can still be found among their descendants and those who followed in their footsteps. Alaska, this writer's home, is the place of "The Last Great Gold Rush," "The Last Frontier," "The Last Great Race," and the place that itself will last and endure forever.

Stampeders whipsaw planks from logs cut at Lake Bennett to build boats to take them to Dawson. The arduous process took time, was unpleasant for sawyer on bottom.

University of Washington Libraries, Special Collections, HEGG 177B

Klondike Nugget tells story of stampede, struggles

It is fitting to start this look at the people of the Gold Rush with the story of Eugene Allen, the editor and publisher of the Klondike Nugget *newspaper in Dawson. His story is detailed in "The Klondike Nugget" by Russell A. Bankson. Bankson's book was published in 1935 by The Caxton Printers, Ltd. Of Caldwell, Idaho. The book no longer is in print, but an occasional copy can be found online. Branson was contacted by Allen who dropped on his desk a bound volume of copies of the newspaper. The author conducted several interviews with the legendary figure, then pored through the newspaper to get a better insight into the man, his fortunes and misfortunes.*

The book details the story of a man who saw and seized opportunity in the Klondike. It provides a look at the hardships of getting to Skagway, getting over the mountain passes, facing the dangers of crossing the lakes and following the rivers to the Klondike. It reveals the shrewd moves made by a man who had no resources beyond an idea, a gift of gab, and determination. It boasts of an editor who knew where his readers stood and fought against actions of corrupt government officials. It mentions the sad outcome of Allen's many ventures, then ends with a hopeful promise but no follow-up, leaving the reader with far more questions than answers.

Eugene Allen, a salesman in the newly-formed Seattle printing company of which his brother Pliny was a part-owner, was a visionary who became one of the most respected men in the Klondike. He founded a popular crusading newspaper, added a personal shopper service, and expanded his circulation department into an international freighting business. As was the case for many of the pioneers of that new frontier, his visions outstripped his abilities and he ended up broke and in disgrace. His legacy, though, built in less than three years of work in the Klondike, lives on as an oft-researched source of Gold Rush history.

In 1897, after learning of the Klondike gold strike, Allen had an all-consuming idea: He would head north to found a newspaper. It was to be called the *Klondike Nugget* and would be the first in the new boom town of Dawson. Allen—along with hundreds of thousands of other people—had been infected with gold fever after seeing headlines screaming, "A TON OF GOLD!" Word that miners had returned from the Klondike with untold wealth in their luggage became known to the world on August 15 when the *SS Excelsior* landed in San Francisco. Two days later, the *SS Portland* docked in Seattle, greeted by swarms of people.

Allen convinced his brother George and Zach Hillman, a printer at the plant where they worked, to join him. For the partners, the Klondike was the end of their rainbow, the place where fortunes awaited, lying there just to be plucked from the sands.

There was one big problem. The trio, all young men, had no resources. They could not even dig up between them enough money to get to Skagway, the point of entry to their destination. Zach was a printer, but they had no experience in publishing a newspaper. In any other era, there would be no chance for them to pursue the dream that formed in their heads.

But Gene was, after all, a salesman. He believed in what they planned to do. They needed equipment for a newspaper plant and they needed cash. Where else should they turn but the owners of the printing company where they worked? Pliny was a part of the management and arranged a meeting with the owners.

"There's big money to be made," Gene told the group.

Before the meeting ended, the trio had the promise of a Pearl pedal-powered printing press to be used for job printing and a flatbed press capable of printing a four-page tabloid size newspaper. They were also given a stake of six hundred dollars in cash. In exchange, the backers wanted and received a half-interest in the newspaper.

George, a teacher in Everett, Washington, eagerly joined the new enterprise. The adventurers met to make plans, poring over catalogs and spending hours making and revising their to-do lists. It was decided that Zach would book passage on the next boat and make his way to Dawson in order to scout the lay of the land and report back as quickly as possible.

Meanwhile, Gene and George went about the task of determining the items they would need to take along to operate the plant. They decided that enough paper, type, ink, stationery, and office supplies to last a full year would be required. There would have to be type cases, a marble composing stone, tables, desks, chairs and stools, and dozens of other items needed to carry out the publishing business.

They grew more and more anxious as they awaited Zach's return from the north. They watched the crowded docks where men milled about, all equally anxious to leave for the land where they expected to find gold glittering in the very streets. The summer faded into fall and fall into winter.

Finally, in January, the scout returned to report on his findings.

The trip would not be easy, Zach made it clear. There were two routes to the goldfields from Skagway, one over a steep pass and the other by way of a longer, but easier grade that would accommodate freight hauled by horses. He reported that the Canadian border was at the top of the pass and authorities demanded that each person have supplies to last a year—a ton of goods to be transported in addition to the printing equipment and related supplies. Prices in Skagway were very high and items limited. Goods should be shipped from Seattle, it was noted.

Critical to their plans was securing dogs to pull sleds over the snow-covered trails. Although there was a run on the animals by other stampeders-to-be, ten nondescript mongrels were found, their only common trait being that they were large. They had to be toughened and trained to pull. George was assigned the task of tending the canines. It quickly became apparent that a great deal of dog food would have to be added to their supply list.

They booked passage with the Puget Sound and Alaska Steamship Company aboard the *City of Seattle* for the men and their freight. But on February 9, 1898, when they arrived with their dogs at the Colman dock, they were told there was no room for the animals. A hasty conference was held and it was decided that Gene and his brother would go on ahead with the equipment. Zach and Ed Brandt, a reporter who joined the outfit after learning of their plans, were to follow with the dogs.

The ship hoisted anchor that night, crammed tight with about 600 passengers, her holds crowded with freight and animals, with excess freight stacked astern.

Allen told Branson that, looking back, "It was a wonder we didn't go to the bottom." There were a third more passengers than the steamer was designed to carry. She rode low in the water due to the amount of freight on board. The concern hit hard when the ship pulled in at Wrangell on Saturday, Feb. 12, and they learned of the sinking of the *Clara Nevada* with the loss of all on board.

Two days later, the passengers were relieved to see Skagway appear in the distance. They disembarked and saw that their supplies were safely placed in a warehouse, then set out to learn about this bustling entryway to the Klondike. It took no time at all to know that one Jefferson Randolph "Soapy" Smith ruled the Skagway underworld. The newcomers found that Smith's henchmen employed many tactics to learn who had money and how to take it from them. Robberies were prevalent and went unpunished because the only law presence was controlled by the Colorado ex-patriot.

Eager to learn more about this place, and more importantly about the rich Klondike area that lay beyond the mountains, the brothers struck up conversations with anyone willing to spend a few minutes sharing with them. They soon began to have a feel for the area. They heard stories of misfortune, but also picked up valuable tips as they waited for the other two members of their party and the dogs to arrive.

An important item that had to be acquired was heavy duty sleds on which to pile their goods for the long trip ahead. Gene slapped down ninety dollars from his dwindling supply of cash to purchase three sleds. On February 17 Zach and Ed arrived with the dogs. The party wasted no time dividing the freight amongst the freight sleds, each pulled by three dogs; the tenth dog added to the sled loaded with the heaviest items.

It soon became apparent that the sleds and dogs could handle only a portion of the freight at one time.

To illustrate the magnitude of their task, an inventory of the printing equipment included the job printing press, a cast-iron contraption weighing about three hundred pounds, the newspaper press, a much larger item weighing almost half a ton, and scores of fonts of type in various sizes. The larger press produced two eleven-by-seventeen-inch pages at a time, the type, made of lead, for each assembled page weighing eighty pounds or more. Type enough for the other two pages plus that of other size letters held in reserve came to perhaps six hundred more pounds. Newsprint for two thousand copies for each of fifty-two weeks was anticipated, plus extra for contingencies, and weighed approximately three tons. Stock for job printing would come to another quarter of a ton. Another quarter of a ton would be represented in incidental material necessary to operate the plant for a year.

In addition to the five-plus tons of newspaper equipment and supplies, each of the six men in the party had to have a ton of provisions—a requirement imposed by Canadian authorities. That included tents, Yukon stove, cooking utensils, boots, a variety of clothing, and food. A suggested grocery list included two hundred pounds of evaporated fruits and vegetables, four hundred pounds of flour, one hundred pounds each of sugar and beans, twenty-five pounds of coffee and five pounds of tea. Matches and first aid supplies were a must.

Stampeders who chose the shorter Chilkoot trail rather than the White Pass route Zach selected had to climb a very steep trail. It started as somewhat easy terrain where animals could be used to move loads. But at Sheep Camp at the base of the mountain, animals were useless because they could not navigate the steep slope. It was four miles from that camp to the summit, the final stage consisting of a 35-degree slope. Goods were carried on the backs of men. Some stampeders hired packers, but most could not afford the wages paid to the Natives who had learned about bargaining—and sometimes sold out to a higher bidder part-way up the mountainside. Once the rush was in full swing, a tramway was erected to haul freight, but few of those who began the trip could afford the tariff. By the fall of 1899, a narrow-gauge railroad had been constructed in a major engineering feat. The White Pass and Yukon Railroad still operates between Skagway and Whitehorse, but now hauls far more tourists than prospectors.

Allen and his group found it necessary to divide the freight into smaller loads, leaving one man behind with the bulk of the freight while three went ahead to establish a cache where the empty sleds were turned around. Leaving a guard at each station forced Allen to employ two other men to help. A break came when the group came across an ill man camped alongside the trail awaiting an empty sled to take him back to Skagway for medical attention. The man had a horse that he could not care for. Allen talked him into letting the newspaper party use the horse, which would be turned over to the ill man's friends when they reached Lake Bennett.

In a full month's time, the Klondike-bound team had covered just twenty miles, but they finally reached the summit with all their gear intact. It was downhill from there to the banks of Lake Bennett, twenty miles away. They had encountered parties who met with disaster. At what became known as Deadhorse Trail they saw overworked horses drop dead in their traces, to be rolled over the edge into a ravine below in order to clear the trail for waiting traffic. The ravine was littered with dead animals. One passerby swore he saw one horse pause at the edge, too tired to continue farther under its heavy load, then leap to its death below. It is estimated that three thousand animals perished after being driven beyond their limits.

Lake Bennett was a beehive, even in the dead of winter. The great Rush not yet in full bloom, the land surrounding the lake was already laid bare of trees. Tents were crowded together near log cribs erected to cut lumber to be used in building boats and rafts.

The Allen party watched workers whipsawing planks from logs placed atop the cribs. One man stood on top, guiding the saw on the upstroke and pushing down when it was his turn. The man on the bottom not only had to work harder pulling the long blade through the green wood, but had to endure the sawdust that rained down into his eyes and down his sweaty back. In late March, the mosquitoes were already out in swarms. They attacked viciously in search of blood and drove both man and beast mad. Allen observed that "many a partnership broke up on the spot."

After crossing the summit, Allen learned that a rival newspaper plant was on its way to Dawson and was camped at Lake Bennett, waiting for the river ice to open. It had come from Juneau under the direction of the publisher of a paper in the older gold camp at the head of the Gastineau Channel. The news was a stunning setback.

It was here that Allen's intuitive sense blossomed. He told the others that he would go ahead to Dawson, leaving them to handle the freight. He learned that the owner of a newly-constructed craft was looking for two experienced boatmen to take them through the rapids and on to the Klondike.

"I'm your man," Allen told the boat owner, and headed out to find someone to fill the second slot. He had occasion to question that hasty move when they had several narrow escapes navigating the treacherous waters that were still carrying chunks of ice.

To the relief of everyone aboard, Allen and the other occupants safely arrived at Klondike City, commonly known as "Lousetown." Allen crossed the bridge linking that unpleasant place with the new boom town that was Dawson. He immediately set about looking for a spot to set up business. Brimming with optimism, he approached the owner of a vacant parcel in the center of the primitive business area. When the newspaperman-to-be finished his sales pitch, he had permission to use that lot as his headquarters. Without spending any money, he soon had lumber to erect a billboard, a tent to use for himself, and permission to use the typewriter in the office of his benefactor.

Not yet seeing signs of the expected competitor, Allen began making his presence known among the business people, the hopefuls who had not yet found work, and the miners. He already had subscription blanks prepared before they left and began getting them signed, charging an ounce of gold per year. He had expected the price would be $10, but quickly found the common legal tender was gold dust. A scale and assorted weights were quickly added to his "office."

While circulating in his sales calls, he made note of happenings. His notebook soon had everything from important developments to tidbits involving various workings on the creeks surrounding the town. In the evening, he typed out short news notes and tacked them onto the bulletin board at the edge of the street. Emblazoned at the top of the billboard was the name, *The Klondike Nugget.* Below that was the line, "Dawson City's first newspaper."

The pleased editor saw lines of people form in front of the billboard, eagerly reading the brief items. He was even more pleased when the other newspaper crew arrived a few days later without fanfare and slowly got set up. By the time the competitor was finally ready to go, Allen already had signed up as subscribers most of the population and had advertising contracts with the established businesses.

By the time George, Zach and Ed arrived with the *Nugget* plant, Gene had found a place to construct a building. The first printed edition was gobbled up by the news-hungry population of the bustling camp. More reporters were hired to gain news. Allen even got the idea to send what he called "the Klondike man" upriver to meet incoming boats, get any newspapers they might carry, and interview the newcomers regarding the latest news from "Outside." His newspaper "scooped" the competition with the latest news—even though it was several weeks old.

Street sales of the weekly editions went briskly, thanks to enterprising newsies, including two senior citizens. One, a sweet-appearing grandmother type, was very enterprising, buying a supply of papers as they came off the press, peddling them and returning to buy more. Her sincere and endearing smile earned her more in tips than did the price of the paper. Sixty-five-year-old "Uncle Andy"

Young went among the throng of men crowding Dawson's main street shouting, "The *Nugget*, the *Nugget*, the dear little *Nugget*." His pitch was so earnest that he sold hundreds of copies in an evening.

Seeing a market among miners who were too busy working their claims to get into town, the enterprising publisher established a delivery service using horses in the summer and dogsleds in the winter. When a miner asked the carrier to pick up something and bring it out to him on the next trip, the obvious next venture was to institute a personal shopper service. That grew into mail delivery and, finally, a freight service over the Pass from Skagway. He soon transported gold shipments and opened a financial transfer service.

Before the end of the first year, Allen bought out his partners and solely owned the newspaper. Business was booming and Allen was becoming wealthy—but only on paper since he was investing his income on expanding the burgeoning enterprise.

While all this was going on, the newspaper was growing in readership. Allen began to notice discontent among the miners over practices of the government representatives. One Thomas Fawcett was assigned to Dawson as gold commissioner. He made the regulations and settled disputes between the miners. Soon, some of the commissioner's actions were brought into question.

The last straw was when Fawcett set a date of July 11 for the opening of claims on benches above one of the creeks. Late in the day of July 8, however, the commissioner posted an amended order, moving the filing date to the very next morning. Advised in advance, several people drove stakes and posted their claims two days ahead of those who were only aware of the original order. It was later claimed that members of Fawcett's staff or his cronies were now the owners of the valuable claims.

Allen ran a series of news articles unfavorable to the commissioner, then followed with editorials calling for Fawcett's removal. George Allen led a delegation to the capital in Ottawa and succeeded in having the gold commissioner replaced.

Not limiting its attacks on Canadian officials, the *Nugget* was fearless in reporting the follies of Dawson's citizens. On one occasion, the newspaper was sued by United States Consul James McCook for an account prominently printed about the American official. According to the story, at half-past three one morning, McCook entered the Phoenix with dance hall queen "Diamond-Tooth Gertie" Lovejoy on his arm. He caught the attention of the house when he offered to buy a drink for anyone who was proud to be an American. Only one patron failed to rush to the bar, that man loudly making known his objection to saluting the Star Spangled Banner. A scuffle of noteworthy proportions ensued. It ended when McCook pulled out a small version of that flag, attached it to his backside, leaned over and invited the objector to give him a swift kick. McCook sued the paper and both of the Allen brothers, asking twenty thousand dollars from each for the pain and humiliation suffered from having the escapade made public for all of the Yukon to see. The suit was dismissed after a long line of witnesses showed up to testify for the defense. McCook soon was recalled by the State Department and left the Yukon.

Gene Allen was hailed as a hero and was felt to be the most powerful man in the Yukon because of his vigorous fights with officialdom.

Meanwhile, large mining companies began to buy up

claims and ordered heavy equipment to go into promising new areas. They planned to also work over "mined-out" claims to recover gold lost or missed by the primitive sluice boxes.

One of those companies contracted with Allen to bring in some big equipment. He promised to get it to Dawson in time to begin work the next season. Delay after delay was encountered, however. Allen felt he had to meet the promised date, regardless of the cost. By the time the shipment was finally on the way and possibly to be delivered on time, he had spent far more than would be received for the freight. Not only that, but an anticipated gold transfer did not come in to pay his financial backer in Seattle. His note was called. Allen was defeated, deeply in debt and without credit.

Meanwhile, miners from the Klondike region were beginning to head downriver to Nome where a big new strike had been reported.

One afternoon, before word of his predicament became known, Allen drew up a paper assigning his interest in the newspaper and its subsidiaries to his attorney. That evening he stopped in many of the bustling saloons and had a friendly drink with his friends. Shortly before midnight, he met up with a man he trusted. Allen had asked the younger man if he'd like to go to Nome to follow the stampede there. Getting an eager affirmative response, he asked to have two sleds and dog teams together with supplies there, ready for the trip. He did not say who the other person would be. The musher was not disappointed when it turned out to be Allen.

Allen told the author of the history that he hoped to strike it rich in Nome and repay his creditors every dime. Whether he did is one of the questions left unanswered.

Allen told Bankson that he planned to start a *Nome Nugg*et and that the equipment ordered for it was lost in a shipwreck. A newspaper by that name was started as one among several on the Seward Peninsula, but was not run by Allen. Allen apparently did not linger in Nome.

Klondike Nugget news items

James Messenger is angry and disgusted and not without good cause. For several weeks he dug and prospected on a vacant bench claim opposite 27 Bonanza but finding nothing but a few colors he abandoned the ground and came to town for a rest. Thursday morning he was chagrined beyond description when friends came in to tell him that the ground had been worked by someone else who had already rocked out $4,000 and was being offered $25,000 for the claim. James has gone out to see for himself, and if it proves true he has a standing invitation by his friends to "come and kick me."

FINED FOR FISHING ON SUNDAY

J. Dugard and P. Fleming thought they would go and examine their fishing nets last Sunday, and now they wish they had let them alone. Fishing as well as labor of any kind is unlawful in Dawson on Sunday, so Messrs. Dugard and Fleming were asked to contribute $2 and $3 costs for the good of the nation.

THE PAVILLION THEATER

Among the new faces that will appear at the Pavillion tonight for the first time in Dawson are the following: Gracie Robinson, Myrtle Drummond, Lucille Elliott. The older, but none the less popular, stars will appear in new acts, among them are the following: Jacqueline and Rosaline, Nellie Lamore, Emma Hull, Eva St. Clair, Annie Kane, Fred Breen, Chas. Brown, and Dick Maurettus. The able management of Mr. Gardner tells in the large attendance at each performance.

Columnist 'Stroller' White wrote of Rush with humor

One of the "*Nugget* men" who was to stay in Alaska was Elmer J. "Stroller" White. A columnist before that designation became fashionable, he frequently urged his readers to "always put a squirt of lemon in it." He was fond of writing about "the Sam Brothers: Flot and Jet."

Born on the family farm in Cambridge, Ohio, on November 28, 1859, when he became old enough to realize that a farmer's life was not for him, he migrated to Florida. There he worked for a brief time as a school teacher, then decided to try his hand at newspapering. He moved to Washington Territory shortly before it became a state. While there he met and married Josephine, who was to remain with him throughout the rest of his life. She, in fact, took over *Stroller's Weekly* in Juneau, Alaska, after her husband's death in 1930.

Although he steered clear of covering politics for the many papers for which he toiled, he served in the Alaska Territorial Legislature and was elected Speaker of the House. He was appointed as director of the Bureau of Publicity and served in that capacity for three or four years before it was abolished by the Legislature. Mount Stroller White at Juneau is named for him.

Although "Stroller" White was, and continues to be,

one of the most quoted newsmen in Alaska, there is little written about him. One of the most complete descriptions of the columnist is in the preface to "Tales of a Klondike Newsman" by Robert A. DeArmond. DeArmond, who himself became an honored Alaska newsman and publisher, as a young man worked with White.

Hearing newsboys shout news of the Klondike strike, White found it impossible to cure the itch that affected his feet, booking passage for Skagway. He wrote of that trip in his first article for the *Skagway News.* The following description is his and could not be improved upon:

> "Thus it was that the Stroller boarded a rickety little wooden steamer one fine morning in Seattle. Passenger space was at a premium and the Stroller felt fortunate to be aboard. That is, he felt fortunate until it was too late for him to turn back. The vessel had been built originally as a tug, and after the start of the Gold Rush she had been remodeled to carry passengers. She was fitted up to take about twenty passengers in fair comfort, the Stroller would judge, and there were seventy-six aboard when she sailed from Seattle. It was her maiden voyage in the passenger trade and it was the Stroller's maiden voyage on the briny deep. The boat had sailed, after her remodeling, without undergoing any trial runs, and so had the Stroller, and the trip was more or less of an experiment all around.
>
> "Just about everything went wrong that could go wrong. The caulkers must have missed a few of the seams and the pumps had to work continuously to keep her from sinking. When the engine was humming along smoothly, the steering gear went on the

> blink, and after it was fixed, the engine pulled a strike. There were head winds all the time and the water didn't stay flat and smooth but kept humping all the time and swishing around in a manner that was very disconcerting to the Stroller and most of the other passengers. The gangplank washed overboard and was lost and so many things happened that the officers began to declare that one of the passengers must be a voodoo."

The voodoo was eventually identified as a certain soup-slurper from Pittsburgh. How he was found out and subsequent events of the trip make interesting reading, but the readers of this account will have to find a rare copy of the book to learn more. Suffice it to quote White as saying, "any man who consumed soup with such a resounding intake would put a jinx on a battleship, let alone a frail, done-over tugboat."

White's biggest story in Skagway was the fatal shooting of "Soapy" Smith by Frank Reid and Reid's subsequent demise due to the confidence man's gunshot.

He left the Alaska port city for the glamor of frontier Dawson across the border in Canada. In Dawson, the Stroller found a great deal about which to write. His columns give an insight into the life of the tent city and the demands on writers of the various newspapers to cover its happenings.

Some weeks there was little going on in the bustling boom town, especially in the wintertime. In such a case, he wrote, if a fire consumed one of the many tents considerable attention would be given to the story. With a large amount of space to be filled in that week's paper, the reporters on the scene would give all the details that could be dredged up: the time the alarm was sounded, whose tent or tents were involved, the names of the fire-fighters who showed up to douse the blaze, a list of the contents of the tent and the extent of the damage. If there still remained space to be filled, the story would include the names of as many spectators as could be ascertained.

On the other hand, if it was a busy news week, any fire that did not threaten the whole town received only bare mention.

White was an accomplished newsman but he preferred looking for the offbeat angle. He said he preferred writing about "The Sam Brothers, Flot and Jet." He sought out local characters rather than dignitaries. They were his kind of people.

In extreme cold weather, the reporters were reluctant to leave the stove that warmed the newspaper office. As deadline approached, the editor began to issue all manner of threats until his staffers headed out to seek news to fill the yawning holes in the pages. White admitted freely that in many such cases, news was not easy to find. A handy device often used in the camps was to find "an old Indian" who could be relied upon to provide enough information to meet the need.

A description of one such day and its outcome was given in one of the Stroller's columns. He and another reporter had remained near the heat until they no longer could risk the editor's wrath leading to violence. The yawning holes were wide indeed and were on Page One, yet. Desperate measures were needed. Unfortunately, White could find no one who could provide anything of interest. Apparently, neither could his colleague. Both returned to the office and sat down to write.

White's story was that a party had just arrived from

downriver and had big news. Noah's Ark, lost since biblical times, had been discovered on the hills above the Koyukuk River. He wrote in great detail about the discovery and how the party was on its way Outside to report to authorities so that an expedition could be arranged to retrieve the hallowed artifact.

The other reporter had found an old Indian who showed him ice worms in a glacier that formed on a hillside not far from town. These creatures were alive and wriggling in the ice, making their way through tiny tunnels that thawed open as the worms wended their way through the blue of the solid ice.

It was decided that the Stroller's story was more newsworthy and deserved the big banner headline at the top of the page. There was one problem, however. The plant's large type only had two of the letter K. "Noah's Ark Found in Koyukuk" was out of the question. The ice worm story made the top position that week due to a slight technicality. White was devastated, according to his account of the event.

Readers of his columns, though, were already accustomed to putting a squirt of lemon in it.

THE SEATTLE POST-INTELLIGENCER.

SEATTLE, WASHINGTON, SATURDAY, JULY 17, 1897. EIGHT-PAGE EDITION.

KLONDYKE

COOPER & LEVY

Store Open Until 9 O'clock Tonight.

COOPER & LEVY

LATEST NEWS FROM THE KLONDIKE.

9 O'CLOCK EDITION.

GOLD! GOLD! GOLD! GOLD!

Sixty-Eight Rich Men on the Steamer Portland.

STACKS OF YELLOW METAL!

Some Have $5,000, Many Have More, and a Few Bring Out $100,000 Each.

THE STEAMER CARRIES $700,000.

Special Tug Chartered by the Post-Intelligencer to Get the News.

BRINGING BACK GOLD.

Pages 13 to 22. The Examiner. — Pages 13 to 22.

REPORTS FROM THE FAR-AWAY LAND WHERE THE EARTH SEEMS LINED WITH GOLD.

TURN THEIR EYES TO THE LAND OF GOLD.

Hundreds of Fortune-Hunters Are Hurrying Northward.

MORE THAN ONE TON OF NUGGETS.

Another Cargo of Gold From the Clondyke Bonanzas.

Publishers jump on news with eye to profit

Coverage of the Klondike gold strike was big news. Publishers jumped on the story, seeing a chance to provide a service and make a few bucks doing so. Even those who had never been there presented "expert" advice on how to go to the Klondike and find success.

A remarkable feat was that of John W. Leonard, who wrote "The Gold Fields of the Klondike." The book was published in a matter of weeks by A. N. Marquis & Company of Chicago. Out of print after its first edition, a copy was found by a Dawson resident and given to George Shaw, a businessman of that city. It was amplified with the addition of photographs and updated with later information and re-published in 1994 by Clairedge of Whitehorse, Yukon Territory. The original cover is subtitled "Fortune Seekers' Guide to the Yukon Region of Alaska and British America," confirmation that it was intended as a how-to book hurriedly put in print to satisfy the demands of a populace anxious to learn about this strange new land and perhaps themselves follow the North Star.

Leonard was there when the North American Transportation and Trading Company's *SS Portland* arrived in Seattle in the middle of July, 1897, with nearly two tons of gold aboard, the property of some of the owners of the first fabulous claims on the Klondike. Leonard interviewed the newly wealthy "Klondike Kings" to get their stories. He started with these, designed to stimulate interest, then proceeded to list supplies needed by those who were expected to join the stampede. The publisher hired artists to illustrate scenes for the book as described by those who had been there or as imagined by the illustrator. A few actual photographs were obtained.

What the book contained is best described by the chapter titles themselves: I-Gold on the Klondike, Wealth Laden Gravel in a new El Dorado. II-The Klondike Discoveries: How Adventurous Pioneers Found Fortune in the River Banks. III-Pioneers of the Klondike: Stories of Those Who Were in the First Rush to the Bonanza District. IV-Joseph Ladue's Story: The Klondike Strike Described by the Owner of Dawson. V-Official Reports: High Authority Confirms the Richness of the Klondike Strike. VI-The Yukon Region: General Description of the Great Gold Fields District. VII-Yukon Gold Country: Mining Developments in Alaska and Northwest Territory. VIII-How To Get There: Who Should Go, and How They Must Outfit and Travel. IX-The Juneau routes: How to Go Down the Yukon from its Headwaters. X-Dawson City: Description of the Metropolis of the Yukon. XI-How Women Fare: Experiences of Mrs. Berry and Mrs. Gage, Told by Themselves. XII-The Climate: Those Who Go to the Klondike Must Prepare for Cold Weather. XIII-Gold and Its Distribution: Quartz and Placers: How Gold Came to the Klondike. XIV-Working Placer Mines: How Gold is Taken from the Ground in the Klondike District. XV-The Law and the Miner: How the Prospector Must Proceed to Get His Claim. XVI-Alaska: Facts About the Land of Ice, Seals and Gold.

Later researchers have found Leonard's reporting to be fairly accurate. His accounts of the stampeders generally reflect their true situation while his instructions on preparing for the trip north fall short of actual requirements. His descriptions of the mining process are correct and are based on his own experience in mining in Australia and at various American locations. Unfortunately, nothing other than that note about his qualifications is to be found in modern archives.

Some Outside publishers got into the act, turning out newspapers with Klondike nameplates but printed in the States. One such was *The Klondike News*, datelined San Francisco, Cal. and Dawson, N.W.T. Volume I, No. 1, shows a copyright date of 1898 by Virgil Moore. Its front page is a gold and black duotone with a full-page drawing of George W. Carmack, listed as "discoverer." The thirty-eight-page inaugural edition of the quarterly consists mostly of feature stories about miners, with a few how-tos and informational tidbits thrown in.

A six-inch article buried on the bottom of Page 16 in the April 1 edition under the head ADIOS apparently was written to fill the last hole left in the pages of the edition. It starts off, "Tonight the Editor sleeps. For full five months he has been in a state of insominous inoxuetude. The first three months of the year he spent in running up and down the creeks seeking the information contained in these pages. Then came six weeks of tramping over the ice and traveling by steamer and rail from the sanctum in Dawson to the composing room at San Francisco, and now we are finished. Hurrah!" The item ends, "Now a little rest and then—"The Summer Edition." It is signed with the single name Valle. No second edition of the *Klondike News* has been found.

Photographers join Rush in effort to record history

In addition to written accounts, much of what we know about the Yukon-Alaska Gold Rush comes from the many photographs taken by hardy artisans who joined the stampede. Hoping to sell souvenirs to the miners, they had little idea that their pictures would be of great value to historians a hundred years in the future.

Theirs was no easy task. Today's cellular phones fit in the palm of a hand and instantaneously capture a high definition image at the push of a button. The digital image can be edited in many different ways and stored for instant recall at any time. Unlimited prints can be made, one after another, several in a minute's time and on a choice of papers...even reproduced on t-shirts, cups, labels or whatever can be put through a digital printer.

Not so in the Eighteen Nineties. Cameras were bulky affairs that had to be placed on a tripod in order to avoid blurriness due to long exposure times. A bellows separated the lens and the plate on which the image was stored. That plate was glass, covered with a silver salts emulsion compound that the photographer carefully applied to the surface in a light-proof room. The light-sensitive plate had to be protected until it was inserted into the camera. Surviving photos from the period are generally of good quality, perfectly exposed and in focus. Both steps were determined by the photographer's eye. Light meters had not yet been invented, meaning that the photographer had to judge the light conditions and time the exposure based on past experience. A black fabric hood covering the back of the camera, the image was examined by the photographer on a ground glass screen fixed into the camera frame. The lens was moved forward or back by extending or retracting the bellows until the image was determined to be sharp. Once the camera was focused, the lens cap was affixed to shut out light and the emulsified plate inserted. To take the photo, the photographer removed the lens cap, counted out the desired number of seconds of exposure, then covered the lens with its cap. An opaque light-proof clip was put back over the glass plate which then was taken to the darkroom to be developed into a negative.

Chemicals used in developing both the negative and prints made from exposing the negatives to light-sensitive paper were affected by temperature. They could not be re-used and were both bulky and heavy. Fortunately, these could be stored at the photographer's studio and did not have to be lugged to where the photographs were taken.

Among the most prolific of Klondike photographers was Swedish-born Eric A. Hegg. As a youngster he had apprenticed as assistant to a photographer in Wisconsin, where his parents settled after coming to America in 1871. At the age of fifteen, he opened his own studio in Washburn, Wisconsin. When he was twenty-one he moved to the Puget Sound area in Washington and by 1897 had two studios in the Bellingham area. Near the end of that year he joined a group of men, one of whom was his brother, Pete, who were headed to the gold fields. Too late to make the voyage from Lake Bennett to Dawson before freeze-up, he and Pete remained in the Dyea and Skagway camps. There they set up studios to photograph the sourdoughs headed Outside and the newcomers who were preparing to head over the passes.

Hegg captured many images of stampeders climbing the "Golden Stairs" of the Chilkoot Trail and those heading out over the White Pass route. He was still at Dyea on Palm Sunday, 1898, when a series of avalanches swept down over the Chilkoot climbers. His photos show the havoc the slides created and the heroic actions of the survivors who worked frantically in an effort to save those who survived. Tragically, sixty-five persons are believed to have been lost in the largest slide. Several of those were workers on the tramway which was being improved. Some bodies were never identified. That tragedy could have been much worse had not many people heeded the warnings of Natives and sourdoughs who recognized the conditions conducive to a slide. There had been heavy snowfall, then warming followed by more snowfall. Indian packers refused to work because of the danger.

The best known of Hegg's photos are views of the Chilkoot stairs where a human string stretches from the staging area at Sheep Camp at the base of the mountain to the gap some two thousand feet above.

In the summer, Hegg moved to Lake Bennett and then to Dawson. He used a cart pulled by a goat to haul his camera and equipment around Dawson. On the side of the cart was a sign advertising "View Scenes by Hegg." In addition to his brother, Hegg was joined by his wife who operated the studio at Skagway. In 1898 he hired an assistant, Per Edward Larss, who had arrived in March and remained with the photographer.

Constant travels took him to all the creeks where miners toiled in their search for gold. He looked for interesting views and photographed the boat-building at Bennett, the assorted flotilla that sailed from there to Dawson, mining operations, cleanups where gold covered the bottom of sluice boxes, social events, aftermath of fires and the buildings gracing the streets of the camps and towns. Those travels were a strain on his marriage and he went through a bitter divorce that saw him turn over the Skagway studio to his wife. She later sold many of the negatives retained in their files; many others were lost or could no longer be credited to Hegg.

When the Dawson rush began to wane, Hegg moved his studio to Nome where he continued to take pictures that recorded the people and landscape of that town. After leaving Nome, he lived for a while in Hawaii then returned to Alaska, living in Cordova, where he worked for the Copper River and Northwestern Railroad until 1918. He returned to Bellingham to work in his old studio, keeping the business until 1946. He died in San Diego, California, in 1955.

Many other photographers recorded images of life in the Yukon during the Gold Rush. The Kinsey brothers, Clarence and Clarke, had a studio at Grand Forks, a bustling community that sprang up outside Dawson at the confluence of the Klondike and Yukon rivers. The brothers hailed from Snoqualmie, Washington, and joined the stampede of 1898. They mined at Gold Hill above Grand Forks, with Clarence spending his time on mining while Clarke concentrated on photography. Their pictures are of very high quality and depict people and scenes of Grand Forks. Many of them appear in the book, "Klondike Lost," by Norm Bolotin.

William E. Meed went with his wife Mabel to Dawson in 1898, where he worked on riverboats. An amateur photographer, he compiled an extensive collection of Gold Rush photographs taken by Hegg, the Kinseys, H. J. Goetzman, H. C. Barley and George G. Cantwell. The collection now is held by the McBride Museum of Yukon History in Whitehorse, Yukon Territory.

Volunteers rally to recover bodies of fellow stampeders who were buried in the Palm Sunday avalanche on April 3, 1898. At least 65 people lost their lives. Photo by Eric A. Hegg
University of Washington Libraries, Special Collections, HEGG 202

Tappan Adney covered Rush, wrote authoritative book

Loading gold on Bonanza Creek, 1898. Tappan Adney photo

Edwin Tappan Adney was dispatched with his pencil, pad, and camera to the Klondike as a journalist with the famed *Harper's Weekly* and was one of the first of his profession to reach the Yukon in 1897. He spent two years there, writing dispatches for the news magazine and freelancing for other publications. A follow-up trip was made to Nome in 1900.

Adney's book, "The Klondike Stampede," was published in 1900.

On the scene when the rush began, he saw first-hand the development of Dawson City and knew the people who became the first to realize the fruits of their labor on the virgin creeks—and the failures endured by many more who were not so lucky. He was there, too, for the winter of 1897-98 when the threat of starvation caused rationing of food sold from the traders' stocks.

Adney was known for human-interest articles. One told of encountering a tourist headed to the Klondike, not as a trader, but "to have a good time." The man's outfit, he reported, consisted of thirty-two pairs of moccasins, a case of pipes, a case of shoes, two Irish setters, a bull pup, and a lawn-tennis set. Even more unusual was the story of forty-seven-year-old Miss Blanche King, en route to St. Michael. Her entourage included a maid, a cook, a horse, a parrot, three canaries, a piano, two Saint Bernard dogs and a sealskin suit.

The journalist traveled around the Klondike, visiting the claims and talking with miners. He learned what was involved in every aspect of their work and wrote of it informatively. His book is filled with the stories of the miners and of the worth of their diggings—but the worth figures apparently came not from owners but from other miners. Adney noted that the claim owners were reluctant to reveal their own financial information, most likely due to the high tax charged by the Canadians.

Also mentioned were stories of odd turns of events. In the early stages of development in the Klondike district, claims were roughly measured. On Bonanza Creek, he wrote, John Jacob Astor Dusel staked Number Two Above. Dick Lowe came along and measured the claim and found that it was around seventy-eight feet too long. Lowe staked and filed on the excess. Offering to sell it for eight hundred dollars, he found no takers. There likewise was no interest in leasing the slice, so he decided to mine it himself. A worker sank a shaft but found nothing. Lowe then burned anther hole and found one of the richest spots on the creek, at least on a per-foot basis.

Despite himself trumpeting the fortunes taken out by the Yukon miners, Adney wrote despairingly of the nationwide headlines that led to the stampede by inexperienced people, some of whom lost their lives. He blamed newspapers hungry to sell advertising to outfitters and, of course, to increase circulation.

Adney was born July 13, 1868, in Athens, Ohio. He did not remain a journalist, but devoted himself to a study of boats, particularly the birch bark canoes of Native Americans. Returning from the Klondike, he married Minnie Bell Sharp of Woodstock, New Brunswick. He enlisted in the Royal Canadian Engineers in 1916 and served as an engineering officer at the Royal Military College of Canada. His duties consisted primarily of creating models of fortifications for training purposes.

Becoming a Canadian citizen in 1917, the artist decorated the lobby of the Hudson's Bay Company store in Winnipeg, Manitoba, with large murals depicting the company's early history. One hundred models of various water craft are on display at the Mariner's Museum in Newport News, Virginia. He is credited with saving the art of building birch bark canoes after spending years studying the art developed by North American Indians.

Poet Robert Service wrote of Klondike from miners' tales

SPELL OF THE YUKON

There's gold, and it's haunting and haunting;
It's luring me on as of old;
Yet it isn't the gold that I'm wanting
So much as just finding the gold.
It's the great, big, broad land 'way up yonder,
It's the forests where silence has lease;
It's the beauty that thrills me with wonder,
It's the stillness that fills me with peace.

One of the best known writers of the Gold Rush, whose works certainly are the most recited, did not even experience first-hand the events of which he wrote. He was captivated by the tales told by miners who were on their way to or from the creeks. His most famous poems, "The Cremation of Sam McGee" and "The Shooting of Dan McGrew," were in fact written while he worked in a bank in Whitehorse, after the rush had died down and before he ever set foot in Dawson.

Robert W. Service was born on January 16, 1874, in Preston, Lancashire, England, as the first of the ten children of a Scottish banker, also named Robert. The family moved to Scotland where the future "Bard of the Yukon" went to high school. At age 6, he was said to have written a blessing he recited on his birthday:

God bless the cakes and bless the jam;
Bless the cheese and the cold boiled ham:
Bless the scones Aunt Jeannie makes,
And save us all from bellyaches. Amen.

When he became 21, Service traveled to Vancouver, British Columbia, then traveled around the West Coast from Mexico and California before returning to British Columbia. During the intervening seven years he held several jobs, even working as a gardener in a bordello, according to historian Pierre Berton, who conducted several hours of radio interviews with the poet.

Hired by the Canadian Bank of Commerce in Victoria in 1903, Service the following year was sent to Whitehorse, in the Yukon Territory. It was there he met Elmer J. "Stroller" White, then the editor of the *Whitehorse Star*. When Service mentioned that he was trying to think of a poem to present at an upcoming church concert, White suggested that he write about the colorful Yukon area.

It was after this suggestion that the banker/poet was inspired to write several poems. The line "A bunch of the boys were whooping it up" struck Service after he passed a saloon where some men were carousing. *The Shooting of Dan McGrew* flowed quickly as he wrote the lines on large sheets of heavy paper tacked on the wall of his cabin. Service liked to listen to the tales of the sourdoughs and a yarn spun by one of them was the basis for *"The Cremation of Sam McGee."* That one, Service was quoted as saying, was done over a single night, completed by five o'clock the next morning.

In 1908, Service compiled some of his verses into a booklet. He sent copies along with a check to cover printing costs to his father, asking to have them published. The publishing house selected by Robert Senior returned the check, instead offering a ten percent royalty on each volume sold. It was said that the typesetters and proofreaders read the poems aloud and their co-workers were so enthusiastic that the publisher strongly felt the book would be a success. *"Songs of a Sourdough"* subsequently went through many printings and was widely circulated.

The following year, Service quit the bank, rented a cabin in Whitehorse and wrote his first novel, *"The Trail of '98,"* which became a best-seller. A wealthy man by 1913, he traveled to Paris and continued to write. When the World War broke out, he attempted to enlist but was rejected. He became an ambulance driver instead. After the war he continued to write, most notably poems published as "Tales of an Ambulance Man." Years later, one of his poems mocked Germany's Adolph Hitler. When the Nazis began moving toward France, Service, his wife Germaine, and their daughter Iris moved to the United States and settled in California. They later learned that the Gestapo had come looking for Service after occupying the town where they lived.

Service was pleased when he was asked to play himself in the movie *"The Spoilers,"* a story of the Nome Gold Rush, even though the poet had not been at that end of the Yukon. He boasted of his part in a scene with actress Marlene Dietrich, who co-starred in the movie with John Wayne and Randolph Scott.

Although contemporary literary critics dismissed Service's poems as "doggerel and imitations of Rudyard Kipling's work," those opinions were not shared by millions of readers. Service captured the feel of the Gold Rush stampeders, who reveled in his humorous tales told in rhyme. Service himself did not style his work as poetry, but merely as verses.

Service returned to France after the war's end and lived there until his death on September 11, 1958. He was buried in Cote-d'Armor, France. Although his wife and daughter visited Whitehorse and Dawson in 1946, Service declined to go along. He said he preferred to remember the place as it was. His cabin remains in Whitehorse where the Canadian park service maintains it as a tourist attraction. Visitors may peer through the windows, but are not allowed to go inside due to the fragile nature of the building and its artifacts.

Still recited today, "The Cremation of Sam McGee" is one of Service's most popular renditions of the Gold Rush.

For a hefty fee, Chilkoot Trail tramway moved heavy items over steep slope. Photo by Eric A. Hegg
University of Washington Libraries, Special Collections, HEGG 335

Jack London survives scurvy, pens popular books

A gifted writer who spent a year in the Northland said he "found" himself in the Klondike. The whole world was to find him three years after he returned to the States and wrote the best-selling novel "Call of the Wild." A touching story of the Gold Rush, featuring a mongrel dog that became a hero, the book brought fame not only to the author but also gave insight into life in the far north.

Born on January 12, 1876, in San Francisco, California, to an unwed mother and a father listed on his birth certificate as John Griffith Chaney, the boy's childhood was an enigma. His mother, Flora Wellman, was a descendant of Puritan settlers and the daughter of the first wife of a prominent Ohio canal builder. She practiced as a spiritualist, claiming to channel with an Indian chief. Moving to San Francisco after her father remarried, Flora was living with William Chaney when she became pregnant. Chaney, an astrologer, packed up and left about the time the baby was born. The couple had been trying to make a living by offering their talents to the public: he by checking his astronomical charts to predict the client's future according to the alignment of the stars at the time of his or her birth, she by going into a trance to contact departed loved ones for answers to her client's questions. Both were popular "sciences" at the time.

Chaney later studied the law and hung out his shingle. His history suggests that his stars were never to be found in a favorable position. He frequently complained of being a victim of circumstances, blaming various factors for preventing him from being successful.

Soon after Chaney abandoned Flora and the baby, she married a disabled Civil War veteran, John London, already the father of two daughters. To avoid confusion, the child who had been called Johnny changed his nickname, adopted his stepfather's surname and thereafter was known as Jack London.

When young Jack entered school, he displayed an eagerness to learn and quickly was able to read. A librarian strongly encouraged him and he devoured the volumes she made available. The boy found a place to do his homework at Heinold's First and Last Chance Saloon near the San Francisco waterfront. There he was enthralled by stories related by sailors who frequented the place. An interest in the sea was sparked by those conversations. He turned that interest into practical, while be it unlawful, experience at the age of thirteen. Borrowing money from the former slave who had been his wet nurse as an infant, he bought a sloop and sailed it around the bay, pirating oysters from traps set by legitimate operators. Apparently recognizing the juvenile's expertise in that field, authorities solved two problems by hiring him to patrol and ward off adult pirates. In so doing, they ended piracy by the juvenile who was causing problems and also used him to help dissuade others from stealing the valuable mollusk. A couple of years later, he signed on with the schooner *Sophie Sutherland* for a seal-harvesting voyage destined for Japan. That trip completed after enduring a typhoon that threatened to send the ship and its crew to the bottom, he returned to find high unemployment resulting from the depression that was being especially felt in the West. He began tramping across the country, finding jobs wherever he could. At one point, he was given thirty days in jail as a vagrant, experiencing there degradation that long haunted him. Discouraged, he returned to the Coast.

After recounting some of his experiences while entertaining family and friends, the then-seventeen-year-old was encouraged to enter a writing contest for amateurs that offered a twenty-five-dollar first prize. Entries were received from a variety of writers, including graduates of universities in and around San Francisco. Despite having only an elementary school education, London captured the top prize. He was inspired by that to write of his adventures. He submitted many short stories and articles to local and national publications. Few were accepted, although editors wrote encouraging letters of rejection.

As had so many others, he recognized opportunity for both adventure and profit when he read of the *Excelsior's* arrival with news of the Klondike strike. He persuaded his step-sister Eliza to finance a trip north. She agreed to mortgage her home, but only on the condition that her husband, Captain James H. Shepard, who was much older than the twenty-one-year-old Jack, go along. Jack quickly read everything he could to learn what to expect and gathered the needed gear and supplies. He also looked up a miner who had returned from the area and sought his advice on what to expect. After they arrived at Dyea, Shepard decided to go back home rather than tackle the challenging Chilkoot Trail.

Jack divided his gear into manageable loads, packing them one at a time up to The Scales, then repeating the process to complete the last steep climb to the summit. He met several men who banded together with him to make the hazardous trip. At Lake Lindeman they felled trees, sawed them into planks and built two boats. With freeze-up fast approaching, the experienced and water-wise London assumed command of the party and hastened to cross Lake LeBarge in order to avoid being caught by freezing temperatures. Being closer to Dawson would be an advantage in avoiding the crowd following behind come break-up. They settled in for the winter in an abandoned mining camp which had a few cabins.

As was the case with the seafarers who visited the San Francisco saloon where London absorbed their tales, he listened to the Yukon old-timers, making notes that would serve him well in the future. One example of his use of others' experiences is clear in his description of the abuse of horses and mules that perished in Deadhorse Gulch. That horrid place was on the White Pass trail out of Skagway. London had crossed the Chilkoot trail out of Dyea, yet his description as quoted elsewhere in this tome is so vivid that one might think he had personally observed the scene.

It was at the winter camp that scurvy made its presence known as a result of the men's diet of bacon, beans and bread. London's gums began to bleed, his teeth became loose, his skin paled and his joints ached. The men pulled part of an old cabin down to make a raft and as soon as the river opened, the severely ill man was taken to Dawson. There he saw Father William Judge, the Catholic priest who had gone there to build a church. Seeing the need for a hospital, the "Saint of Dawson" established that, too, and tended to the sick and injured. London was fed raw potatoes to ward off the scurvy. The priest advised the young writer to return to the States as soon as possible in order to receive medical care.

Recuperating in Dawson, London rented a room from brothers Louis and Marshall Bond. Both were mining engineers, educated at Yale and Stanford. Both also were Republicans while London was becoming a strong socialist. The three carried on long discussions, debating their widely differing viewpoints; their positions were strongly defended, but always on friendly terms. The Bonds' half collie, half St. Bernard mix would later be the basis for Buck, the major character in "Call of the Wild."

Having spent the summer at Dawson and absorbing the boom town's flavor, London obtained a skiff that he and a companion rowed downriver. The trip was leisurely as the writer viewed the scenery and examined the people and activities in the villages and mining camps along the way.

Although still suffering the effects of the scurvy when reaching St Michael, London obtained passage on a steamship bound for San Francisco, working his way by shoveling coal into the boilers below deck. When he was too weak to work, his shipmates covered for him to allow him to rest before returning to the hot, dusty work.

At some point during 1897, London contacted Chaney, his supposed birth father. Chaney denied paternity, declaring that he was impotent, saying therefore Jack could not be his son. Most of London's chroniclers believe he was, in fact, the father of the author even though Flora was something of a free spirit herself. Chaney was living with Flora for more than a year before the child was born; the man also was known to have been a womanizer who had several "marriages." Chaney died shortly before "Call of the Wild" was published. Had he an inkling of his son's pending worth, his recognition of parenthood likely might have been on the positive side.

London married Elizabeth "Bessie" Maddern, a long-time friend, on April 7, 1900. Their union was not framed by love, but the belief that they were compatible and would produce strong, intelligent children. They made their home in Piedmont, California, where Jack wrote "The Call of the Wild." Their first daughter, Joan, was born on January 15, 1901, and their second, Becky, on October 20, 1902. Jack took pride in his children but sought companionship outside his wife's bedroom—a fact that was not kept a secret from her but caused both resentment and concern over the consequences of unsafe sex. They became separated in July of 1903 and a divorce was granted in November of the following year.

Although a few short stories he wrote had been published, London did not gain wide notice until "Call of the Wild" appeared. It was first published in *Saturday Evening Post,* which paid him seven hundred and fifty dollars. It was picked up by McMillan, who launched a major national advertising campaign. The first printing of ten thousand copies was sold out immediately. In the end, more than a million copies were sold. Unfortunately for London, he sold the rights to the book for a flat sum of two thousand dollars. His fame was established, however, and there was great demand for his works thereafter.

A prolific writer, he made it a point to produce a thousand words a day, working at his desk early in the morning. One biographer claimed that London "had an uncanny knack of knowing when he reached a thousand words, often stopping in the middle of ..."

In 1905, London married Charmain Kittredge, the secretary of McMillan publisher George Platt Brett, Senior. Wanting children, they were denied when one child died shortly after birth and a second pregnancy ended in a miscarriage. Concentrating on their home, he added more acreage and began building a stone mansion called Wolf House. Two weeks before they were to move in, the house was destroyed by fire. They continued to live in the expanded ranch house, frequently traveling to Hawaii.

They returned to the ranch in July of 1916 with Jack in poor health. The effects of a life of hard drinking had caught up with him. He was said to have had his first taste of wine at age three and two years later was drinking hard liquor. His grade-school studies at the First and Last Chance Saloon were accompanied by steins of beer. Throughout his adult life his companions were astounded at the amount of liquor he could hold. Too much of it, it turned out, was held in his liver. He died on November 22, 1916, on the sleeping porch of the ranch house, the result of uremic poisoning. His ashes were buried on the ranch following a private funeral service attended by only a few friends and relatives. Charmain's ashes were buried nearby following her death in 1955.

The grave is marked only by a moss-covered boulder. It is located near the still-standing ruins of Wolf House. The site is now Jack London State Historical Park in Glen Ellen, California. It is visited frequently by fans of the author. London wrote twenty-three novels, scores of short stories, essays and poems. He also wrote two autobiographies. One of those was "John Barleycorn" which detailed his battle with alcohol and called for people to practice abstinence from liquor.

On January 12, 1998 the Friends of Libraries U.S.A. added Heinold's First and Last Chance Saloon to the Literary Landmarks Register.

Stampeders sought many routes on rush to riches

Various routes were used by the anxious gold-seekers who rushed to the Klondike after learning of the discovery of gold. Even the fastest of them found all the ground near the fabulous early discoveries already staked by prospectors who were then in the area. Unfortunately, that word had not reached the ears of the main body of stampeders who rushed northward in the latter part of 1897 and during 1898. Neither had they heard about the hazards to be faced in getting there.

Many Canadian patriots sought to find an overland route across the breadth of their country. They found the prairies easy to navigate, but when they neared the far western portion of their nation it was a different story. The Canadian Rockies and rushing streams in uncharted territory proved to be formidable barriers. The few who were successful in reaching their destination were often in desperate straits due to the extended time their travel required. Historians generally blame selfish Canadian interests for tragedies that befell those who took that advice.

Some sought a shorter route through Alaska, hoping to save time by sailing into a more southerly port, then traversing overland to the Yukon. They failed, however, to reckon with the vast terrain in their path. Many lost their lives attempting to navigate what they mistakenly thought to be shortcuts. What was erroneously expected to be smooth glacial ice turned out to contain deep and threatening crevasses and ridges that rose to heights difficult to scale. Rivers and mountains along the mostly unmapped vastness proved insurmountable to travelers.

Two ports proved to be the safest and most popular entries to the Yukon River and the goldfields along its nearly two thousand-mile northerly flow from headwaters in Canada to where it emptied into Norton Sound.

The shortest ocean route from the West Coast was via the Inside Passage, hugging the North American coast from Puget Sound through Lynn Canal to the Alaskan port at Skagway. For a time it appeared that the more southerly and already settled ports of Wrangell and Juneau might provide access, but the long overland leg and difficult terrain proved unsuitable. A trading post at Dyea was at the foot of Chilkoot Pass, a route used by the Chilkat Indians and recently opened to others—for a price. Nearby Skagway was at the foot of White Pass. There initially was no dock at Skagway, but passengers and cargo were transferred to smaller boats and taken to shore. Wharves were quickly built, however, allowing boats to tie up and unload passengers and freight.

Dyea landing on Taiya Inlet consisted of relatively flat land; five miles inland was the foot of a steep mountain reaching thirty-five hundred feet above sea level. At the base was Sheep Camp, which became a staging area for the ton of supplies each miner was required to possess before entering Canada. Enterprising early arrivals cut steps in the packed snow to make it easier to trudge up the steep grade. A toll was charged of each traveler for use of the steps—a fee gladly paid when viewing the alternative of an unbroken trail through deep snow. Later entrepreneurs built a tramway, but the heavy fee charged for hauling freight in buckets dangling from the cable was too much for most of the newcomers. What became known as the Chilkoot Trail had been closely guarded by the Tlingit Indians until the stampeders arrived in force. The Natives quickly learned that their services as packers brought good compensation. Their rates varied based on demand; at times the packers even abandoned an agreement with one party to switch their services to another who offered higher pay.

Newcomers making the ascent packaged their loads in duffel bags averaging fifty to sixty pounds in weight. With an individual's ton of personal items, the bare minimum requirement of each person resulted in about 40 round trips. Sheep Camp became overburdened with discarded non-essentials as disheartened people rearranged their priorities and their packs after seeing the task ahead. At The Scales resting place was a flat area where packs were re-weighed because rates were hiked from that point on. Between that point and the summit the elevation increased by one hundred fifty feet, on a thirty-three degree slope. Berton quoted one traveler who said that section was so steep it "was like scaling the wall of a house."

The other route from Skagway was via the White Pass, some six hundred feet lower in elevation and at a less severe grade although in places the trail followed some very narrow ledges. The grade allowed animals to carry or pull loads, making it easier on travelers who could afford to buy or hire beasts of burden.

Packers who abused their livestock caused an estimated three thousand horses, mules and oxen to lose their lives on that trail. Many packers were inexperienced in loading as well as lacking knowledge of how to care for the creatures. In many cases the animals were aged and already should have been turned out to pasture. Hay and oats were expensive, in short supply, and forage was limited. Even when food was available, the packers' greed was so great that no time was given for the animals to rest and be nourished. Many collapsed under their loads, slipped off the trails to perish in the canyons or fell through the ice to drown under their loads. One stretch had so many carcasses that it was known as Deadhorse Canyon, making travelers heartsick at the sight.

Twenty-one-year-old author Jack London, who was to become famous for novels written after his 1897 trip to the Klondike, wrote graphically of the situation:

> The horses died like mosquitoes in the first frost and from Skagway to Bennett they rotted in heaps. They died at the rocks, they were poisoned at the summit, and they starved at the lakes; they fell off the trail, what there was of it, and they went through it; in the river they drowned under their loads or were smashed to pieces against the boulders; they snapped their legs in the crevices and broke their backs falling backwards with their packs; in the sloughs they sank from fright or smothered in the slime; and they were disemboweled in the bogs where the corduroy logs turned end up in the mud; men shot them, worked them to death and when they were gone, went back to the beach and bought more. Some did not bother to shoot them, stripping the saddles off and the shoes and leaving them where they fell. Their hearts turned to stone—those which did not break—and they became beasts, the men on the Dead Horse Trail.

Tappan Adney, covering the Gold Rush for *Harper's Weekly,* wrote in his book, "The Klondike Stampede," "Frequent users of both trails said they were equally bad. Whichever one was taken, the traveler wished he had taken the other."

Lake Bennett straddles the border between British Columbia and Yukon Territory, at the foot of the north slope of the Coast Mountains which host both Chilkoot and White passes. The boundary separating Alaska and Canada is near the crest of the mountains. Called the gateway to the Klondike region, the lake is about thirty-three miles from the Alaskan port of Skagway. The lake was named by Lieutenant Frederick Schwatka of the United States Army. An explorer who mapped part of Alaska following the 1867 purchase from Russia, his 1883 trip was sponsored by James Gordon Bennett, Junior, editor of the *New York Herald.* Schwatka's mission was to search for the remains of the ill-fated Franklin Expedition which visited Alaska in 1878 through 1881.

The lakes and rivers were iced-in from early October until late April or May. Travel could be made on foot or by dog team, but extremely cold temperatures and high winds limited the number who ventured between Skagway and Dawson during most of the winter.

The Bennett lakeshore became a large settlement from the winter months of 1897 through the first lustrum of the Twentieth Century. Its prominence faltered when the White Pass and Yukon Railroad was put through to that point in 1899. The railroad was later extended to Whitehorse. In May of 1898, the Mounted Police counted eight hundred fifty-two boats under construction at Lake Bennett and another twelve hundred were thought built over the next couple of years.

Trees on the mountainsides around the lake were felled to provide lumber for vessels needed to traverse the water route that would lead the successful boatmen hundreds of miles north and west. Most of the stampeders had no experience in building boats or navigating treacherous waters. Little did they know what lay ahead of them. Those arriving in the latter part of 1898—those who could afford it—were able to buy planks from enterprising businessmen who set up saw mills to cater to the horde of would-be miners. For most, though, the cost was beyond their means. Of necessity they were forced to rely on their own initiative as had the ones who arrived before the sawyers set up shop. As the haphazard flotilla grew, the available timber rapidly disappeared.

The newly-minted boat builders had no pattern and no directions for constructing watercraft. The result was a *pot pourri* of plans for rafts, skiffs and barges. It required brute strength to fell trees and drag or roll them from where they were cut to the lakeside construction site. Cutting the logs into planks was a two-person job using long whip-saws. An elevated framework was built to hold the logs. One sawyer stood on top astride the log while the other stood on the ground below—where he was showered with sawdust that was sticky with sap. The one on top guided the saw along a marked line while the person below pulled hard on the down-stroke. The spruce logs were green, making the sawing an arduous task. The saw's teeth were far apart and needed to be filed frequently to keep the blade sharp.

A boat with a prow was most favored, based on advice from people who had experienced such a trek previously. Those were much easier to steer through the waiting rapids than was a square-prowed craft. The conveyance's caulking, sturdiness and maneuverability would be fully tested in the days following its launch into the icy waters. That they were in a "rush" was increasingly evident as the signs of spring began to show. It was a race to get the craft ready and loaded and to get underway. Everyone knew that land would be grabbed by the hundreds of people waiting to get started. Word had spread that the "good" ground was already staked. All were in a dither to get there.

Lake Bennett was twenty-six miles in length. At its end was Caribou Crossing which connects Bennett with Tagish Lake. If there were a breeze, a sail could help move the craft toward its destination. If the breeze turned into wind, the lake surface formed waves that threatened to swamp the craft. If there were ice, it could cause damage. So, too, would a collision with a rock or submerged log. Sailors who reached Windy Arm found it was aptly named. Still another river presented its challenges before emptying into Marsh Lake. Covering a distance of twenty miles, Marsh Lake is so shallow and swampy that some who crossed called it "Mud" Lake. At its end is the Lewes River, a winding stream with many pitfalls and narrow canyons along its twenty-six-mile course. There the river entered Miles Canyon, where rushing waters following steep cliffs in many twisting turns caused boaters to refer to it as "Dead Man's Canyon." Exiting that stretch, the boaters had two miles of relative relaxation before suddenly coming upon Whitehorse Rapids. Those who didn't get swept into the raging caldron and could pull into the left shore were able to portage boat and cargo over a narrow, slippery and rocky two and a half miles before casting off again. In the fifteen miles before they reached Lake LeBarge, they navigated Squaw Rapids, not nearly as fearsome as the previous white water travails.

Robert Kirk, an Englishman who made the trip in early 1898, traveled over the Chilkoot Trail and built a boat at Lake Bennett. In "Twelve Months in the Klondike" he wrote, "Almost every party that reached Dawson had a tale to tell of narrow escapes from capsizing on the lakes, and it is probable that not a few fatalities of which nothing is known occurred from this cause."

Kirk gave a first-hand account of the experience:

"Miles Canyon is three-quarters of a mile long, with perpendicular walls...through which water rushes with great speed...the rush of water is so great that it crossed the pool (a large pool lay in the middle of the canyon, with strong currents on each side) in long waves, forming small whirlpools on either side. It became necessary for the travelers to guide their boats through the dangerous place and the speed of the water is such that the run can be made in about four minutes. The run is not considered to be so very dangerous if the boat can be kept in the middle of the stream but several casualties have already occurred. Several boats were wrecked against the walls of the canyon at the time we were there."

It is impossible to obtain accurate information regarding the number of deaths by drowning in these rapids, but this has been estimated at fifty, a considerable portion of which occurred during the spring and summer of 1898.

Lake LeBarge offered a scenic spot to rest before continuing on their journey. Actually a widening of the Yukon, the lake is thirty-one-and-a-half miles long and at its widest, its banks are three-and-a-half miles apart. Its waters are extremely cold and the ice is the last to go out of any of the lakes in the region. Its vista is magnificent enough to have caused inspiration for Jack London to mention it in two of his books and for Robert Service's famous poem, "Cremation of Sam McGee."

At the northern end of Lake LeBarge is Thirty-Mile River, a narrow, zig-zag, rock-filled waterway that proved more dangerous than any of the streams so far encountered. It connected with Big Salmon River and thirty-six miles later with Little Salmon River. That led to Five Fingers Lake and then to a stretch of the river known as Rink Rapids. Fortunately, that was the last of the dangerous waterways. And for those who survived, it was near their goal at Dawson. They had covered nearly five hundred miles of the roughest terrain and water that anyone could have imagined.

Many travelers, however, did not make it. Some turned back when they lacked the required provisions to get past the Mountie checkpoint and did not have the money to acquire them.

Hundreds died along the route, including at least sixty-five people who lost their lives on Palm Sunday, 1898, in a series of avalanches on Chilkoot Trail, below The Scales. Heavy snow was falling on that April Third day and a warm wind was blowing. Native packers and old-timers warned of the danger but their cautions went unheeded by the many who were in a rush to reach their destination. Workers who were constructing a tramway designed to speed the transport of goods up the incline were urged by their bosses to stay on the job in order to complete the line as quickly as possible. Most of them were to die.

There were actually five Palm Sunday slides, three of them deadly. The first occurred about two o'clock in the morning, burying twenty people near The Scales. About nine-thirty, three others were buried when a second section of overhanging snow gave way; in this case all three were rescued. About eleven o'clock three people were crushed in their tent. An ox named Marc Hanna was buried nearby but was extricated unhurt.

The deadliest slide was still to come, worsened by an unfortunate turn of events. A party of tramway workers was the first to evacuate. Visibility was very poor due to the heavy, blowing snow and they took a turn leading into a ravine some distance from the trail. They were lost when overwhelmed by rushing snow and ice. About two hundred and twenty people followed them, although having the foresight to take along a two-hundred-foot rope onto which they held in an effort to stay together. Unaware of what had happened to the tramway workers, they followed in their footsteps. The largest and most destructive of the series of avalanches descended on them after seventy or more had entered the ravine. The front group was buried under as much as fifty feet of snow. The people at Dyea heard the roar of that huge slide but were not aware of the tragedy until a survivor came panting into the camp. An alarm of gunshots was raised and the population turned out to aid in the rescue and recovery of victims. Marc Hanna, the ox that had been rescued, was pressed into service hauling corpses to Dyea.

The actual number of casualties is unknown as various lists were contradictory. The remains of some victims were buried in the Dyea cemetery while others were shipped Outside. Published lists vary. Names appearing on one are not on others; many on those lists do not match names on the cemetery headstones. It is generally felt that more than sixty-five people perished on that day while two others were killed in a slide the following day.

The lakes and rivers claimed many lives in both winter and summer months. Dog teams pulling loaded sleds and accompanied by their owners fell through the ice and were drowned. Some attempting to travel on foot either suffered frozen limbs or succumbed to the cold. In the summer, heavily loaded craft slammed into rocks or swamped in the raging rapids. Inexperienced helmsmen were unable to control their vessels when encountering extremely swift currents. The journals of the stampeders who made the trek are filled with vivid descriptions of watching rafts being swamped or battered on the rocks in the treacherous rapids. There were many more stories of daring rescues, the names of the heroic rescuers unfortunately escaping mention.

Just to safely complete the trip from Skagway to Dawson, no matter what the season, was a feat. That so many did was a testament to their stamina, as well as their determination. While the majority of those who made the trek through to the Klondike were men, many women also made the trip. Rather than live up to the reputation of being "the weaker sex," many observers were of the opinion that they stood up better than the men. There are several instances where historians note that husband-and-wife teams split up during the climb over the Chilkoot, the man giving up and returning home while the woman went on. In one such instance, a couple was planning to open a store. The man gave up, but the woman went on with their goods to establish the business and fared well.

A look at the dangers presented along the route to the

Klondike is presented in an obituary from the first edition of the *Caribou* Sun, a newspaper that sprang up in the settlement on the shores of Lake Bennett. Luc Richard's name was misspelled as "Lue" in the account published on May 10, 1898, six days after the event.

THE CARIBOU SUN

VOL. I. CARIBOU CROSSING, (HEAD WATERS YUKON RIVER,) N. W. T., MONDAY, MAY 10TH 1898 NO. 1.

TWO MEN DROWNED

Lue Richard and Thomas A. Barnes meet With Death in Lake Bennett

Last Tuesday a serious accident took place near the Island about midway of Lake Bennett which resulted in the death by drowning of Lue Richard and Thos. A. Barnes. They with O. S. Felton and H. M. Buck were on their way to Bennett with Richard's dog team for supplies. When near the island the party became alarmed at the poor condition of the ice and started toward the shore when the ice gave way; Richard at once sank to the bottom but Barnes hung on to the ice for about ten minutes and made a gallant struggle for life, but sank for the last time when the rescueing party had almost reached him. The accident was seen from the shore by a number of Yukoners who rendered all aid possible and succeeded in rescueing Messrs. Felton and Buck, who lay flat on a small patch of white ice which barely sustained them. The dog team and sled were lost. All valuables on the bodies were recovered. The body of Mr. Richard was recovered the next day, but the searching party did not find the remains of Mr. Barnes until the day following. A reward of fifty dollars was paid for the recovery of the bodies.

The funeral took place on Friday afternoon when Mr. C. A. Walsh read the Episcopal burial service and a choir sang Rock of Ages and Nearer My God to Thee. A very large attendance of men and a number of ladies were present and contributed a profusion of wild flowers. The burial took place on the island about forty feet from the water line among a garden of wild roses. Headstones were placed on the graves properly inscribed and a picket fence will enclose the Island's first cemetery.

Mr. Richard was about thirty-eight years of age, of French descent, and came here from Frenchtown, Montana, in company with Will P. Brayton, Mike Beaulieu and Charles Bouchard also of Montana. Thos. A. Barnes was about thirty-five years of age, an Englishman by birth, whose residence was in Axtell, Kan. He was a member of the Iowa-Alaska Mining company who are in camp here.

White Pass & Yukon Railway an engineering feat

In little over a year, the stampeders' tortuous climb over the Chilkoot and White Pass trails would be replaced by a four-hour train ride. The cost would be just half of the toll charged by the man who placed a toll gate across the access to the White Pass trail. Eliminated also would be the "Golden Stairs" of the Chilkoot Trail, the backlog of goods cached along the trails as stampeders went back and forth with fifty- or sixty-pound loads on their back, and the livelihoods of professional packers.

So reported Cy Warman, who wrote a series of articles on railroads for McClure's Magazine. His eight-page "Building a Railroad into the Klondike" article appeared in the March, 1900, edition of the illustrated monthly.

Warman visited Skagway in August, 1899, and interviewed the president of the White Pass and Yukon Railroad, Samuel H. Graves, and his chief engineer, E. C. Hawkins. He rode the train to Lake Bennett just one month after the first passengers made the trip and only fifteen months from the time Hawkins and his assistant, John Hislop, began looking for a route where the rails could be laid.

As unusual as the short construction time was the start of the railroad company itself. Seeing the potential, several groups put forth plans to build a railroad connecting Skagway with the Klondike. One of them, upon receiving a permit to proceed, borrowed money from Close Brothers financial group in London, a financial institution still in operation today. One condition was that they complete the road within a certain period. When they were unable to do so, the bank elected to proceed with the project on its own—without ever seeing the area and not knowing the challenges that lay ahead. They ordered Graves, then overseeing a large reservoir project in the American West, to go to Skagway and take over the railroad project. Graves took along his top two engineers, Hawkins and Hislop.

The two engineers set out to explore the route. They chose to follow the mountainside rather than go along the river trail used by the stampeders for the first portion of the trip. Trudging along the steep slopes in extreme weather conditions, the surveyors had several narrow escapes with death.

When plans for the railroad became known, opposition from several sources was expressed. Objections by the packers and the man whose toll road was threatened were anticipated, but the threats and protestations of legitimate businessmen were not expected. The merchants and gamblers were upset at the thought that the shorter time newcomers would spend in Skagway would rob them of revenue and they raised their voices against the project. The contractor negotiated for a considerable period but saw little softening in the opposition.

When Graves applied for permission to run track along the outskirts of town to connect with a branch along Broadway leading to the docks, he found the authorities uncooperative and stalling. A public meeting was called to discuss the issue. Debate was heated and lasted until three o'clock the next morning. It ended with Graves getting approval to run the tracks through downtown. The precedent for quick action was set that very moment. When the stores opened the next morning, residents were startled to see a locomotive chugging back and forth along tracks running through Broadway Street.

Graves found it was not difficult to secure workers among the throng that had become disheartened upon finding themselves unable to get beyond Skagway. Many could not meet the requirement of the Canadians that each person crossing the border possess a year's provisions. Most of them had come to Skagway hoping to get rich and lacked the wherewithal to sustain themselves, much less buy a return ticket home. They signed up to work for the railroad at thirty cents an hour. A laborer's pay at the mines was fifteen dollars a day—if only one could get there. Hugh Foy, an experienced railroad builder hired as superintendent, was to find that his crew included many college graduates and professional men as well as clerks, deckhands, farmers, teamsters, cowboys, and former soldiers.

Handed picks, axes and shovels, the workers were sent out to follow the stakes the engineers had placed, clearing away vegetation along the path. Drillers poked holes in the rock faces of the steep mountain to make way for blasters to insert dynamite, then all retreated to safety before the explosives were set off. At that point a crew of laborers dropped to the trail below to clear the path for people and animals struggling through on their way to the goldfields. The drillers far above were already at work poking new holes and the process was repeated time and time again, around the clock. At that latitude, there was no need for artificial light during the summer. By midsummer of 1898, a thousand men were hard at work, moving the track farther and farther north.

The work was hard and it was dangerous. Drillers swung from ropes dropping down from anchors high above. Strong winds hindered their efforts. Even today, scrub trees near the summit are bent, showing the effects of high winds like those of more than a century ago.

Workers clearing debris and smoothing the railbed strained to keep up with the onrushing crews as ties were put in place and tracklayers drove spikes. There was a constant state of urgency and work proceeded regardless of weather. Under such conditions, accidents were unavoidable. In one tragedy, a house-sized chunk of rock fell from above, crushing two men working on the trail below. With no hope of rescuing the workers and the rock of such great size, it was decided to leave it there to become the largest headstone any of them would ever have as a memorial.

There were several serious interruptions. One occurred when there was a rumor of a new gold strike at Atlin in British Columbia, southeast of Skagway. A good many of the railroad workers took off to follow the new stampede—some of them taking along the railroad's tools to use in scooping up the gold they expected to find there. As had the latecomers to the Yukon, they found most of the good ground already staked.

Another delay came when the lead crews reached the Canadian border. They were barred until the company's right to build in that nation had been clarified. The weight of the powerful British owners soon gained that clarification. The White Pass summit, inside Canada, was reached by the tracks on Monday, February 20, 1899, and became the cause of a large celebratory banquet. Foy, the aging and very tired superintendent, attended the banquet and was honored. Shortly afterward, he caught a bad cold which quickly developed into pneumonia. He had been urged by Graves to take time off to rest and was scheduled to leave on vacation on the last day of February. He died that very day, however, and his remains sailed in the hold of the *Rosalie,* the same ship on which his stateroom had been booked.

Still another delay came in December when, due to darkness, storms and cold weather, the company cut workers' hours from eleven to nine per day, accompanied by a pay raise from thirty to thirty-five cents an hour. Workers were upset because the forty-five cents resulting from the increase in hourly pay did not offset the sixty cents lost by the reduction in hours. The same forces that earlier sought to block the railroad's permits recognized another chance to stop, or at least slow progress, came into play again.

Egged on by the selfish interests of Skagway packers and storekeepers, the miners went on strike. After missing their first paychecks, they grew quite angry. The most vocal of the anti-management workers began urging the crowd to advance on the railroad shops, threatening to burn them unless their demands were met.

Hislop, the engineer, was meeting at the time with workers, urging them to take a more reasonable stand. He sent a warning to Heney, the contractor, who was at the shops a few blocks away. Heney rolled a locomotive out of the building, blocking the roadway, and turned on its headlights to illuminate the area. When the angry strikers reached the shops, they hesitated when they saw the blockade. Steadfast in front was Heney, backed by a group of supporters carrying clubs and rifles. One of those with a rifle was the railroad doctor, F. B. Whiting. Hoping to rally his backers, the leader of the strikers confronted Heney in a menacing manner. Whiting grabbed the barrel of the weapon he was holding and clubbed the striker over the head. The miners, most of whom were along mostly out of curiosity, quickly walked back toward the comfort of the saloons. The doctor cast away the rifle, which now had a broken stock, and carried the unconscious would-be arsonist into the building where several stitches were taken to close the wound.

Labor troubles averted, permits cleared, the summit having been reached, and spring renewing promise, work on the tracks began with renewed fervor. Rails were laid along the route until the crews came to a point where a bridge was required. The track-layers then moved to the other shore and went on from there while the bridge-builders filled in the gaps.

From the McClure article by Warman, who rode the train to Lake Bennett just a month after the first passengers boarded the train there, we learn of that historic event:

Early in the morning of July 6, 1899, a steamer brought two hundred passengers from the Klondike to Lake Bennett where they expected to hike over the Chilkoot with their possessions. They were surprised to see a small building with a sign that read, "WP&YRR TICKET OFFICE." Passengers could see no tracks, but a ticket agent was at the window, smiling in anticipation.

"When is the next train," asked one man.

"Leaves at 2 p.m.," was the reply.

"Where does it leave from," was the follow-up from the man, who now had a crowd standing around to hear the answer and expecting it to be one that would draw derision.

"Leaves from this station a 2 p.m.," the agent smiled.

The incredulous passengers looked in the distance. Nothing was in sight up the cleared right-of-way. Suddenly they heard the rhythmic ping-ping-ping of hammers against metal. Within minutes men lugging heavy lengths of rail came into view, other men ahead of them dropping wooden ties onto the ground.

A run was made on the ticket window as the passengers clamored for tickets and tossed their pokes onto the window ledge. Soon a locomotive pulled up. Workers continued laying track, a steady stream unloading ties, spikes and rails, methodically building a loop around which the locomotive could begin the return trip to Skagway, today's mission complete.

The White Pass and Yukon Railroad continues to make that trip today. Skagway is a major stop for tourists cruising the Inside Passage. The locomotives have been upgraded but some of the original parlor cars, or replicas thereof, stand ready to take on passengers. The original stationhouse is now a museum where tickets can be purchased. Each passenger coach has a tour guide who points out the spots of historic significance and describes the difficulties experienced by the workers. From the train, one can see portions of the original White Pass trail along the Skagway River. The tour guide points out the rock covering the remains of the workers who lost their lives and is a memorial to them and their comrades who met their Maker during the project.

During the Second World War the United States Army operated the narrow-gauge railway, hauling supplies from the port to crews building the Al-Can Highway. As was the construction of the railway, that wartime project also was an exceptional feat completed in record time.

A highway now connects Skagway with Whitehorse, Yukon Territory. Today's tourist train stops at the Canadian customs house where tour passengers can return to the cruise ships by way of comfortable buses. Even with a brief stop at a "Gold Rush" show at a trading post well-stocked with souvenirs and a chance to pan out a few specks of gold, the quick bus ride gives little hint of the travails encountered by those who hiked over the passes in 1898, before machines made it an almost forgotten memory.

Passengers on first trip of White Pass and Yukon Railroad at summit of White Pass pause for a picture on February 20, 1899. Photo by Eric A. Hegg

University of Washington Libraries, Special Collections, No. 335

Balloon proposed for access, Aeronaut puts on shows

Before the railroad was constructed, an easier way to get to the goldfields was the object of several enterprising individuals. Word spread that climbing the passes was an arduous task. Many had ideas to solve the problem.

One such entrepreneur announced plans to have balloons float between Skagway and the Klondike. The craft he was building, he claimed, would carry a ton of supplies, making the trip in just two weeks. Although he had many inquiries from would-be passengers, and even one advance payment of five hundred dollars, the project failed to get off the ground.

One balloonist did make a big splash in the Yukon, though, drawing a huge crowd to witness the first-ever balloon ascension in the territory. The *Dawson Daily News* reported the feat of "Professor" John Leonard, Aeronaut, which "attracted thousands of people to the water front" at suppertime one summer evening. The balloon rose slowly from the ground. A trapeze hung below the helium bag, with the aeronaut performing various tricks to the delight of the crowd. Reaching an altitude estimated at five hundred feet, Leonard, wearing a parachute, leapt from his perch and floated toward earth. He then released the parachute and dropped into the icy water.

The daredevil aeronaut gave several other such exhibitions, passing the hat among the crowd and, based on reports, did very well in the way of compensation.

Leonard did not fare so well, though, in an ascension reported by the Dawson paper on September 9, 1899. The show started well, the description best told by the awe-struck reporter:

"Up, up, and away soared the great balloon until its ascensive power began to wane and then, while hanging by the ankles alone—a performance that for skill and reckless daring outrivals the most awe-inspiring feats of ancient Rome—Leonard made the pulse-thrilling plunge with the parachute." His awe then replaced by pity, the reporter said the parachute safely lowered Leonard to the roof of the North American Trading and Transportation Company warehouse only to see Leonard fall from the roof and sustain "a severe strain, from the effects of which he will be laid up for some time." The remainder of the aeronaut's shows that year were cancelled.

Leonard returned in 1900, however, but found the crowds had thinned considerably due to the declining population. He packed his equipment and went downriver to Nome where he entertained good crowds. Nome's airspace was even less forgiving, however, and he was blown out to sea on two occasions.

The idea of aerial transportation did not pan out, but Leonard found that people would pay to watch him soar, perform trapeze stunts, then cast himself off trusting to the canopy that brought him gently to the earth—when not getting dunked due to a wayward wind. Aerial photographs from the period indicate that photographers also were among his customers.

On Canadian side, Mounted Police keep order

While lawlessness ran rampant in Skagway after Jefferson Smith took over the town, the North West Mounted Police ruled the Klondike with an iron hand, but in a fair manner. Created in 1873, the force was well acquainted with the territory, but not fully prepared for the stampede to come in 1897-98. A contingent of Mounted Police was sent to the Yukon in June of 1895, under the command of Inspector Charles Constantine. When the rush of people showed up in 1898, Superintendent Samuel Benfield Steele was sent to supervise the force. He was to be policeman, magistrate, temporary gold commissioner, and at one point in charge of rationing supplies to the population during the threat of starvation. Assigned the duty of border control and collecting custom, in one year the Mounties registered twenty-two thousand people crossing from the United States into Canada by way of the Chilkoot and White passes. They also for a time were asked to carry and distribute the mail—a task that was beyond the capacity of even these intrepid minions of the law.

Staffed mostly by young men, the Mounties earned a reputation for honesty and compassion. At the lakes where stampeders stopped to build boats and start the final leg of their journey toward the Klondike, the distinctly clad men offered kindly advice. When too many accidents occurred in the rapids, or when the ice was not safe for travel, they promulgated and enforced regulations designed to protect the novices from their own greed. Women were not permitted to ride the hand-made and often overloaded boats through the rapids but were ordered to take to the trails high above the river.

One oft-repeated story illustrating the honesty of the force was the trip where a detail of four men was assigned to transport five tons of government-owned gold ingots through American territory and by steamer to a bank in British Columbia. The salary for the men was one dollar and twenty-five cents a day. The fortune in gold was delivered safely to its destination.

The sturdiness of the men who wore the uniform of the Mounties was monumental. At their posts at the top of the mountains, they were on duty through the worst weather that the north could throw at them. With limited staffing, they put in long shifts in extreme weather, registering travelers, inventorying goods and collecting custom. There was no respite from temperatures that reached sixty below and lower. On the trail the care of their dogs came first. When a traveler was reported to be in trouble, the Mounties went to their aid without regard to their own safety.

Steele was strict in enforcing the laws, but recognized that discretion was called for, especially when the preponderance of the residents of Dawson and the surrounding creeks were not Canadian citizens. These men, far from home and hearth and accustomed to other laws, needed to be treated with some leniency. There also was the fact that the governments of the adjoining nations were having some disagreement over boundaries in the northern area; unrest among the residents could boil over and cause an international incident.

Prostitution was illegal in Canada, yet it ran rampant in Skagway and many unattached females had crossed over the passes. When logged in at the border, most of them listed their occupation as "actress." To deny them entry or to deport them on suspicion of illegal acts might well lead to insurrection among the miners. The law enforcement commander, who was also both judge and jury, relegated the prostitutes to specific districts walled off from polite society. Many of the painted women chose to set up business across the river from Dawson in Klondike City, commonly known to the miners as "Lousetown."

"Blue" laws prohibiting the sale of liquor, gambling or work on Sundays were strictly enforced. Crimes such as carrying sidearms, fraud, cheating at cards, disturbing the peace, using foul language, operating unsanitary facilities or rustling of dogs were vigorously prosecuted. One of two possible punishments was assigned: a "blue ticket" expelling the offender from the territory, or hard labor on the woodpile that fed the stoves at the Mounties' facilities. Failure to leave expeditiously with the blue ticket was cause for six months on the woodpile.

Enforcement of the punishment was considered a deterrent and not taken lightly. In one case a gambler accused of taking advantage of a customer was fined fifty dollars by Steele. The insolent cheater snarled, "Is that all? I have that much in my vest pocket." Steele did not hesitate before adding, "and sixty days on the woodpile. Do you have that in your vest pocket, too?"

The hard line apparently was effective. There was not

a single murder in Dawson in 1898 and almost no major thefts even though cabin doors were left open and gold sat in plain sight in cabins whose owners were away. Outside Dawson, two men who were headed Outside disappeared after snow had fallen at the end of the season. Mounties were assigned to investigate. They found a trail that branched off the main route and spotted something leaning against a tree that probably was linked to the missing men. Inquiries about travelers who might have been on the trail at the same time turned up two persons of interest. After a long investigation, the Mounties found the body of one of the missing men in the river. They tracked down and arrested one of the suspects who was found with gold in a poke that was believed to belong to the murdered men. The second suspect was said to have committed suicide. The other victim was never found. The case lent strength to the motto, "The Mounties always get their man."

The Mounties were disliked by many of the packers. Although designated as "mounted" police, few of them rode horses in the Klondike. They respected the animals, however, and had a deep disdain for packers who overloaded, starved and beat the beasts of burden on which their livelihood depended. When an animal with open wounds was spotted, its owner was ordered to care for it until the wounds were healed. If the animal was too far gone or the packer refused to comply, the animal was shot to put it out of its misery. It was widely known that packers would stop before going through a checkpoint and carefully cover any defects that might cause the Mounties to stop them. Writers of Klondike history are almost unanimous in their repulsion at the way the animals were treated. Many of the beasts were not suited to the task to which they were put. The animals were not properly fed. Their taskmasters, themselves often inexperienced, were utterly cruel in the way their animals were loaded—ineptly and bearing far too much weight—and driven to desperate lengths.

On the other hand, the residents along the creeks and in the settlements on the Canadian side of the border appreciated the security given by the presence of the Mounties. While some Americans objected to the taxes collected or the enforcement of rules on provisions to be carried, there was little complaint.

Cy Warman, an American writing for McClure's Magazine, highly complimented the crimson-coated lawmen. He wrote, "At Log Cabin, we greeted the North West Mounted Police. They are the customs officers here, but the most obliging, gentlemanly customs officers you will find in many years' travel."

Many a hapless group of stampeders found it necessary to dry out belongings after their craft swamped on the rivers.

Photo by Eric A. Hegg
University of Washington Libraries, Special Collections, HEGG 719A

Belinda Mulrooney a strong-willed woman

A woman whose business acumen was a match for any other of the Gold Rush's entrepreneurs was Belinda Mulrooney. She was born in 1872 in Ireland but immigrated with her family to the United States as a very young child. She was raised on a farm near Scranton, Pennsylvania, but her aim was for business success.

At the age of 20 she traveled to Chicago and opened a sandwich stand during the World's Columbian Exposition. At the conclusion of that event she left for San Francisco and with her profits from the Chicago enterprise opened an ice cream parlor. Even though the country was in the throes of a depression, the shop was a success and was expanded. Her dreams of a comfortable life were dashed, though, when fire destroyed the building.

Not deterred by disaster, she went to work as a stewardess on a ship operated by the Pacific Coast Steamship Company. The vessel operated between San Francisco and Alaska, calling at various ports including Juneau, where the Treadwell mine was producing gold. On her first trip she realized there was money to be made by catering to the needs of the women of the town. She invested in an inventory of clothing, bolts of cloth and luxuries that would appeal to the ladies. While the ship unloaded passengers, she trundled her goods to town and did a brisk business.

Having saved the tidy sum of five thousand dollars by 1897 when news of the Klondike strike reached Juneau, she recognized that there would be a great many people headed there. Intuitively, she also recognized that those going over the passes would take with them only the bare necessities. She invested most of her earnings in more cloth, more clothing and something she realized would be of value in the north: hot water bottles. She had the goods hauled over the Chilkoot Pass and down to Dawson. There she quickly sold the inventory at a reported six times its cost to her—double today's triple-net theory.

It was said that when she stepped off the boat for the first time, she threw a half dollar into the Yukon, saying it would be the last time she would need such small change. She was to live to see that come true quickly and abundantly.

With the profits from the merchandise she brought to the new gold camp, she built a restaurant in Dawson, then turned her attention to building a hotel and restaurant. Rather than in the bustling boom town, though, she built the Grand Forks Hotel closer to the claims at the confluence of the Klondike and Yukon rivers. Its rooms, restaurant and saloon were popular. Her customers talked freely and she quickly became familiar with the successes and failures of the prospects. By the end of her first year, she owned or was a partner in several valuable claims.

Belinda sold the hotel for a very good price just when gold was discovered on benches above the rich Bonanza claims. She then turned her attention to building what she wanted to be the finest hotel in Dawson.

As described by Pierre Berton in "Klondike Fever," the proprietor oversaw every detail of the planning and even went to Skagway to supervise the hauling of the material over the passes and on through to Dawson. Going there turned out to be to her advantage. When she arrived at the port city she found that Joe Brooks, the packer she hired to transport all the freight, had dumped her property alongside the trail two miles out and signed on instead to pack a load of whiskey for another outfit that offered more money.

Belinda hired a gang of men, staged a fight to determine which among them was the toughest, and set out to catch up with the double-dealing scoundrel who had betrayed her. When she caught up with the offending party, her men tied up the packer in a tent, dumped the other saloon-keeper's load of whiskey, then loaded the hotel merchandise on the prisoner's mules. She saw her furnishings delivered safely in Dawson.

Berton described the Fairview Hotel plans: "It was to have twenty-two steam-heated rooms, a side entrance for ladies, electric lights (with power supplied by a yacht anchored in the river), and Turkish baths. The tables were spread with linen, sterling silver, and bone china. In the lobby an orchestra played chamber music." It took a year to put on all the finishing touches. Among the furnishings were cut-glass chandeliers and brass beds.

The hotel's opening was a success, the dining room, bar and hotel taking in six thousand dollars the first day. Dawson's most successful legitimate businesswoman once formed a partnership with "Klondike King" Alex McDonald to salvage cargo from a small ship that ended up on a sand bar. The mining magnate, though, got to the wreck first and took all the food. Belinda was left with a supply of gum boots and liquor. The liquor she took back to the hotel to be added to her inventory. The boots she held onto. Later, when McDonald needed to buy boots for his miners, she was able to supply them. She charged her former partner a hundred dollars a pair.

A hard-nosed manager of her many affairs, she was engaged by the bank to operate the failed Gold Run Mining Company. The operation's managers, it turned out, had paid more attention to their gambling operation than to the mining business. Belinda got rid of the gambling appliances, drove away the girls sharing the men's camp and imposed strict rules. In less than eighteen months, the bank was pleased to find that the operation was again in the black.

As shrewd as the Irish-born businesswoman was, she made one seriously wrong move. She fell for a dapper dude with a thin black mustache over his upper lip, from below which issued words smoothly spoken with a French accent. Introducing himself as M. Le Compte Carbonneau, a monocle stuck in one eye, spats on his shoes, and kid gloves on his hands, he was accompanied by a valet. Declaring himself to be a favored member of the French royal court, he was there to take orders for fine champagne to be served to her customers. He also was there to woo the richest woman in the Klondike. She fell for his line after weeks of being presented daily with a fresh red rose. Even though a respected miner from Montreal identified the spirits-seller as a barber from that city and an outright imposter, the twenty-eight-year-old spinster in October of 1900 became the "Countess" Carbonneau.

Accounts of their marriage differ. There is agreement that they spent their honeymoon in France. Some say that the “count” was a military purchasing agent during the First World War and died when a German artillery shell struck while he was making a call near the front lines. Others say that Carbonneau skipped out with Belinda’s furs and jewelry and left her high and dry to deal with a raft of litigation.

However the marriage ended, Belinda found her fortune dwindling as the Dawson boom came to an end. As did so many of the Klondike miners, she sailed downriver to Tanana in the fall of 1904. Her business instincts intact, she soon had an interest in claims on three rich streams. She operated the mines profitably, then went Outside. She brought her sisters Margaret and Nellie the next spring to settle in Dome City, a gold camp that sprang up near the site of Felix Pedro’s gold discovery a few miles from what was to become Fairbanks.

Seeing an opportunity, she opened the Dome City Bank in 1906. Named as partners in the bank were Belinda’s sister Margaret and Jesse Noble. Noble married the other sister in January of 1907, but Belinda suspected he was a fortune-hunter and booted him from the bank and threw his trunk out of Nellie’s window.

Having prospered, Belinda and Margaret left Alaska. Nellie remarried and stayed behind. Noble was to become a successful Fairbanks citizen on his own accord.

With the knowledge gained when she built the finest hotel in Dawson at the height of its glory days, Belinda bought a ranch in Yakima, Washington. Perhaps with the lingering memory of her life as a “countess,” the landmark became known as Mrs. Carbonneau’s Castle. With her money at last running out, she sold the ranch and moved to Seattle where she lived quietly. Mrs. Charles Edward Carbonneau died in 1967 at the age of 95.

Harriet 'Ma' Pullen became beloved hotelier

One of the most successful people connected with the rush to cash in on Klondike gold did not make her fortune at the Yukon claims but in Skagway. Harriet Pullen arrived there in 1897 with the first wave of stampeders and remained for the next half century. She became the owner and hostess at Alaska's finest hotel and in later years acted as official greeter of tourists and keeper of the area's history. One account has it that she also kept up the grave of the notorious "Soapy" Smith, a man she did not hold in the highest regard. The story was that an anonymous person sent her fifty dollars each year to see that the infamous Jeff Smith's grave was maintained.

Born August 13, 1860, in Hope, Wisconsin, Harriet Matilda Smith migrated west to Washington, which became the forty-second state seven years before George Carmack filed his discovery on Rabbit Creek. She was 20 when she married Daniel Webster Pullen at Clallem, Washington. They established a horse farm and planted fruit orchards at Cape Flattery. They also had three red-headed sons and a daughter in quick succession.

With the nation in the throes of depression, the man whose given name was the same as that of the famed United States Senator, oratorical whiz, and future secretary of state under three different presidents, skipped out on his wife. She was left with mounting bills and four very young children. By 1897 she had lost everything except the children and seven horses.

On the advice of a friend, the destitute woman left her children and the livestock with another friend in Seattle and scraped up money for passage to Skagway. There was not yet a dock, so she joined other passengers who climbed into small boats to be taken ashore. With only seven dollars in her purse, she was wondering what to do when a man approached and asked if she could cook.

Her first job began that very night, cooking for the crew that was starting construction of a pier to accommodate the horde that was to come in search of gold. Her employer was Captain William Moore, the founder of the town of Skagway.

Finding her three-dollar-per-day salary insufficient to meet her needs, she made extra money by baking pies from dried apples, selling them to anyone who had the money. The pies turned out to be very popular and within a year, Harriet had the money to send for her three sons. Skagway was no place for a young girl, so the daughter remained in Seattle. Harriet also arranged to have the horses loaded on the ship as cargo. It had early become apparent that a great deal of freight bound for the Klondike would be unloaded onto the new pier. She realized that all the heavy freight could not be packed on the backs of hopefuls laboring up the steep Chilkoot Trail. Her horses would be valuable.

Quitting her job as cook, the woman who was to become known and loved as "Ma" Pullen rented a log cabin where she set about baking pies by night. During the day she hitched up four of the horses and hauled freight over the White Pass trail to the summit.

When Captain Moore's family tired of the wild Alaska life and moved "Outside," their stately home stood empty. The budding entrepreneur rented it and converted it to a hotel. She soon bought it and made it into the finest hotel in the Territory. Its rooms, which boasted of bathtubs and soft beds, were always full. Its dining room featured grand china and silver, set on linen tablecloths. The food was superb and there were always apple pies available as dessert. Fresh milk and butter came from Jerseys and Holsteins on the Pullen farm established near Dyea.

The red-headed sons, Dan, Royal and Chester, benefitted from their experience and the teachings of their mother. Sadly, the youngest, Chester, lost his life when his ship went down as he was returning to classes at the University of Washington. Dan was accepted into West Point, where he had a brilliant football career and became an officer. Royal also served as an Army officer. Dan was awarded the Distinguished Service Cross as an Engineer for service during the Great War. Royal, a captain in the Engineers, was also decorated. It was said that General John J. Pershing commented that he wished he "had a full regiment of Pullens" in order to bring a quicker end to the conflict.

The Pullen House remained in operation until Harriet passed away on August 9, 1947. She is buried nearby. The hotel fell into disrepair and was demolished. Its stone chimney remained as a landmark.

Pullen had several competitors in the freighting business. Some of them were henchmen of Soapy Smith. Clients of theirs were lucky to make it into Canada with any of their goods.

Among the others were Charlie O'Brien, a Nova Scotian, who one day walked on board a southbound ship with fifteen thousand Canadian dollars stuffed in his pockets. He was bound for Victoria, British Columbia, to make a deposit at his bank.

Another, Joe Brooks, kept the side pockets of his jacket stuffed with gold coins. It was his habit to walk into Clancy's saloon and slap a stack of coins on the bar and call for everyone to have a drink on him. The portly and obstreperous Brooks was said to have brought seventeen mules with him to Skagway, pushing them off the ship and swimming them to shore. His herd reportedly swelled to three hundred and thirty-five animals and made for him as much as five thousand dollars a day. He is mentioned, unkindly, in the previous look at Belinda Mulrooney, another highly successful businesswoman of the Gold Rush.

Clarence J. and Ethel Berry among successful miners

The lives of many early-day miners became entwined so tightly they are difficult to separate. Such is the case for the first miner to hit the paystreak on bedrock that proved the value of the Klondike strike. A California fruit farmer whose enterprise failed in the depression of 1893, Clarence Jesse Berry the next year went to the Yukon in hope of recouping his losses. After three fruitless years of tramping the tundra, including one period of three months when he existed solely on a supply of beans, he took a job as a bartender at Fortymile, a Yukon settlement inside Canada near the Alaska border.

Far from being discouraged at not finding the pot of gold, Berry was bolstered by the amount of the mineral being taken from the creeks near the Fortymile camp. He had managed to accumulate enough of the yellow stuff to buy transportation back to California. There, on May 10, 1895, he married his childhood sweetheart, Ethel Dane, and convinced her to go north with him. He also talked Henry, the middle one of his three brothers, into going along. The grueling trip up and over the Chilkoot trail without doubt was not the honeymoon of which the young woman had dreamed. Her husband later said, though, that she stood up well under very trying circumstances and did not complain. She dressed in men's trousers and flannel shirts and pulled her weight with the men. When they reached Fortymile, while the men were away visiting the surrounding creeks in search of gold, she lived alone in the one-room cabin her groom had built. The small cabin had no floor, no windows and at first not even a door. Its furnishings consisted of a bed, two roughly made chairs, a stove, and a box that served as a table as well as a storage cabinet.

Still without finding a good prospect (exploratory panning that turned up more than a few cents worth of gold per pan) in order to support his new wife over the winter, CJ took a job as bartender. The saloon in that small mining camp on the banks of the Yukon was owned by Joseph Ladue. The swamper who washed dishes, swept the floor and cleaned the spittoons, was a down-and-out prospector named William Gates. A customer was a handsome young Austrian by the name of Antone Stander, who had staked a claim on a tributary of Rabbit Creek but had no money to sustain himself while digging an exploratory shaft.

All were in the saloon on the fateful day in August of 1896 when George Carmack stopped to order a drink in celebration of his discovery of gold fifty miles upriver. In the spirit of prospectors of that day, Carmack talked freely of his find. There was no hint of the assertion later made that it was his wife Kate and her brother, a Tagish Indian named Skookum Jim, who had actually made the discovery while Carmack fished. Earlier in the day, Carmack had filed his claim on Rabbit Creek with North West Mounted Police Inspector Constantine, acting gold commissioner, renaming the creek as Bonanza Creek.

As usually happened when a prospector told of finding gold, those who had sought but not found anything more than a few tantalizing colors, made tracks for the greener pasture—or in this case, golden creek.

Berry, who was better known by his initials, "CJ," was grubstaked by trader Jack McQuesten and prepared to head for the Klondike, leaving his wife and brother behind. During the night, Gates also started upriver. Stander told Berry that he had filed for five hundred feet on a nearby creek, but that he could not get credit from the trading post to buy supplies. The former manager of the Alaska Commercial Company store had been free with credit. His replacement, however, trusted no one and earned the sarcastically disparaging nickname of "The Pope."

Let us first follow the story of CJ Berry, arguably the most successful of all the Klondike miners. He is the only one of this group of soon-to-be millionaires whose fortune was to be well shepherded, grow, and bring lasting respect to its owner. The others will be mentioned later in more detail.

Berry, who proved to be a generous and considerate man, agreed to co-sign for Stander's supplies in exchange for a half-interest in the young immigrant's claim on the un-named tributary of Bonanza, formerly Rabbit Creek. He promised to also swap a half interest in a claim he would later stake in the Klondike region. Just how well that arrangement was to work out could not have been imagined when the deal was struck.

The Berry brothers poled a small boat upriver against the strong Yukon current, struggling for several days before reaching the soon-to-be famous site fifty miles away from their Fortymile cabin.

They set to work digging a shaft down to bedrock—a

depth not yet known. What they did know from CJ's previous experience was that gold's weight makes it settle through the gravels until it can go no farther upon coming to rest on bedrock, either a "hardpan" of tightly packed material or solid rock. They also knew that streambeds changed over time and that gold was found in "paystreaks" that meandered according to the flow of water in ancient times. The phrase "gold is where you find it" was and continues to be very literally true.

The process generally used in developing a prospect was to thaw the permanently-frozen ground and dig a shaft. This required cutting trees, chopping them into firewood and building a fire in a hole dug from the surface until permafrost was encountered. Necessity dictated that the hole was only large enough for a man to climb through, carrying pick and shovel—usually six to eight feet square. A bucket with a rope tied to the handle was lowered to hoist the gravel and muck from the hole to be deposited on the surface. When the hole was deeper than a man's reach, a windlass was employed to raise the heavily laden bucket. At this latitude it was not necessary to shore up the sides of the hole as the permafrost was solid, with little danger of a cave-in. The Berry brothers were only able to remove about six inches of material before repeating the thawing process. It was tedious work, requiring a great deal of firewood and constant tending of the fire. Firewood was cut while waiting for the ground to thaw, nourishment and catnaps taken whenever there was a chance. The material in the buckets was checked occasionally for signs of gold. When these began to show up as the hole grew deeper, the material pulled out thereafter was placed in a separate heap to be run through a sluicebox when water became available. After a prospect was proven to show value, the shafts were dug during the winter months and the sluicing was done during the summer. If a paystreak were found, workers could be hired to expand the number of shafts.

As it turned out, the Berry brothers were very lucky. Their first shaft hit paydirt. That claim proved to be the richest on Eldorado—soon to be termed the richest creek in the world. A single pan taken from the first shovelful at bedrock was said to disclose two hundred dollars in gold. More than twice that amount was gleaned from the full shovelful. The Berrys were to become wealthy beyond belief.

That winter, however, CJ had the scare of his life. Surveyor William Ogilvie had laid out Joseph Ladue's townsite of Dawson at the hub of the Klondike streams. That task completed, he was begged by a group of miners to verify the boundaries of the many claims that had been located along the major creeks.

A straightforward by-the-books engineer, Ogilvie reluctantly took on the task—but only under the condition that his decisions were final. He had already expressed dismay at the haphazard manner in which the hodgepodge of prospectors had marked the corners of their claims. Under Canadian law, a person who discovered gold could mark out and then file on a claim that was five hundred feet in length, its width going from rim to rim of the streambed. Their boundaries were marked by mounds of rocks or stakes driven into the ground. Measurements were taken by the prospectors without benefit of instruments. Every means imaginable was used to determine five hundred feet, from laying out lengths of rope to stepping off the distance.

In one case a former Mountie had carefully measured

the meandering edge of the stream. By stepping off the winding riverbank, he was cheating himself. Ogilvie found that the claim, by properly using a straight line from the first stake to a point five hundred feet distant, was short by a considerable distance, in fact some seventy feet.

In another case that was loudly disputed and figured largely in the decision to ask for the survey, a man named Jim White argued that two adjoining claims were each longer than allowed. He drove stakes inside their bounds and proclaimed that he was entitled to file on the space he alleged was over the limits. Ogilvie had a suspicion that White was only hoping to get a cash settlement from the two owners and had no intention of mining. He put that parcel off to the last. When the survey was complete, he found that White was correct. There indeed was a gap that was open. It measured three inches in width. White was told he could stake it, but the man at first declined, hoping to find a promising area where he could file the one claim to which prospectors were entitled. When he was not able to locate another, he mined the three inches and found his claim to be the "richest by square inch" of the entire Klondike. From that point on, the obnoxious man was known as "Three Inch" White.

Ogilvie could have made himself wealthy by using his findings to his own benefit, but was scrupulous, law-abiding and professional—in every case but one. Late one afternoon he approached the highly respected Berry and summoned the miner aside so he could speak to him in private. The Berry claim, *Five Above* on Eldorado, the surveyor solemnly confided to the owner, was forty-one feet six inches beyond the five hundred foot limit. Even worse news was the fact that the material piled up from the winter's digging was on the opposite end from the legal starting point, and subject to be claimed by someone else. In shock, Berry asked aloud what could be done to avoid catastrophe. Ogilvie swallowed, then asked quietly, "Do you have a friend you can trust?"

Berton, the Klondike historian, wrote that Berry invited the surveyor to stay for supper, ushered him into the small cabin and asked Ethel to prepare a meal for their guest. Berry then slipped out the door and rushed five miles upstream to a claim being worked by his good friend George Byrne. They dashed back to Five Eldorado where Byrne quite innocently asked the surveyor about the procedures for staking and filing a claim. In the dark of night and out of sight of the workers, Byrne staked forty-one feet six inches below the end stake of *Five*, then signed a lease to Berry for the fraction now owned by Byrne once he filed it with the commissioner.

The mound of gold-bearing material was washed through sluice boxes during spring break-up and CJ and Ethel Berry prepared to go Outside with the fruits of their labors. They were joined by Thomas Libby* and his wife Salome, owners of a claim located a mile upstream, as they traveled to St. Michael. There they boarded the ancient steamer *Portland*. They were among those aboard when the ship arrived in Seattle a month later and were greeted by throngs of reporters eager for descriptions of the Klondike. The Libbys, who were Seattle residents, were even more sought after by the Seattle papers for a local angle to the stories.

After the Klondike rush faded, the Berrys moved downriver and to Esther, near Fairbanks, and staked claims there. While there, CJ developed two different mechanical means of thawing ground, which was extremely valuable in the Interior of Alaska where bedrock was deeper and the permafrost thicker. Berry continued to operate mines in Alaska until 1914. A little more than a decade after he first went north, Berry returned to California where he invested in oilfields. He established Berry Petroleum, naming different fields for members of his family; the Ethel D is still producing, according to the company Web site. Berry Petroleum continues as a wholly-owned subsidiary of Linn Company. He is remembered in the company's history as a kindly and benevolent man.

Clarence Berry died suddenly in 1930 of a burst appendix. Ethel lived a life of luxury in Beverly Hills, California, until she died in 1948.

*Thomas Libby was among a party surveying the Alaska coastline while searching for a site for a proposed underwater cable connecting North America with Europe. He found gold while prospecting in the area, but did not develop a claim. After the cable project was abandoned, he tried his hand in the Klondike.

Antone Stander finds gold, loses all to dancehall gal

Antone Stander was a tall, handsome young man who arrived in New York City from his native Austria in 1887. He was twenty-nine when he reached the Klondike in the summer of 1896. He joined up with four partners who prospected around the Yukon. They had observed activity on Rabbit Creek and moved on to a small tributary of that creek where they panned and found some color. The oldest of the quartet was Jay Whipple, a prospector who had been in the north for several years without hitting a good prospect. Frank Keller had been a railroad brakeman in California. J. J. Clements was from New York. The fourth was Frank Phiscator, a farmer from Michigan. Each of the five staked claims on what was to become Eldorado Creek. Each of their claims was to produce a million dollars or more—but of them only Stander was to enjoy any of that wealth, and that not for long.

Stander went to Fortymile in search of a grubstake, hoping to buy supplies on credit from the Alaska Commercial Company. He had heard that the manager, Jack McQuesten, was a trusting soul who was generous with credit as he had been at his former post at Circle. When the hopeful prospector arrived to make his needs known, however, he was turned down flat by the new manager,

Edgar Mizner. Stiff and unfriendly, Mizner demanded that a guarantor co-sign Stander's promise to pay.

Stander's benefactor was Clarence Berry, whose kindness made both of them wealthy. The two worked their claims and cleaned up their sluiceboxes the following spring.

Carrying one hundred thirty thousand dollars in gold, Berry and his wife were passengers on the *Excelsior* when it brought the word's attention to the Klondike strike. Still single, Stander went looking for a wife, but in the new town of Dawson. At the Gold Hill Hotel, he spotted Violet Raymond, a dance hall girl spirited away from a Juneau entertainment palace by Max Edelman, owner of the hotel. Violet considered herself owned by Edelman and rebuffed Stander's amorous attention. The young miner was so enamored of her that he would not give up. For starters, after hearing that diamonds are a girls' best friend, he bought up every diamond that could be located in the fledgling settlement. Not sufficiently impressed but apparently sensing there was value in playing hard to get, Violet said "no" while continuing to tease. Stander presented her with a necklace of nuggets that was long enough to reach the knees he so admired. Still rebuffed, he gave Violet twenty thousand dollars in gold and promised her another thousand dollars each month. With that, she told Max goodbye and moved in with Stander.

The couple traveled to San Francisco and made plans to go to China for an exotic honeymoon. When the ship tied up at the pier in San Francisco Bay, Violet wanted to go ashore to buy new clothes and otherwise prepare for the trip. That caused a problem for Stander, who had brought along nearly two hundred thousand dollars in gold to use as spending money. Not trusting the purser, he chose to keep the money in his room rather than put it in the ship's safe. Unrealized by her suitor was that Violet intended to spend the gold as quickly as possible. With her bridegroom afraid to leave his gold unguarded in the *Humboldt's* best stateroom to go shopping with his beloved, he doled out a few ounces at a time and watched her go out on her on. Before long, there was no gold left, which caused Violet to also leave.

Stander had invested part of his money in building the Stander Hotel in Seattle. He also put a large amount into alcoholic beverages—not with the intent of stocking the hotel bar but for the purpose of forgetting his lost love. The hotel, located at Fourth and Marion, had two hundred and fifty rooms divided into suites with private bathrooms, telephone service and modern ventilation. What it lacked was a way out for its owner.

One of the Klondike's first millionaires, Stander died a pauper, the victim of alcoholism. Violet sued him for divorce. In the settlement, they divided a fortune of nine hundred thousand dollars. She died in 1944, leaving an estate worth fifty thousand dollars.

Stander married Bessie Longworth in 1909, but they were separated at the time of his death. Broke, he moved into the Pioneer's Home in Sitka but was reported to have been evicted due to drunkenness. He died on April 21, 1952, in Multnomah. Oregon.

Alexander McDonald, a Nova Scotian of Scottish descent, was Dawson's richest citizen based on land ownership. He made a fortune from leases and mining, contributing heavily to the Catholic Church and other causes. He was awarded the Order of Saint Gregory by Pope Leo XIII, the badge proudly worn in this portrait. Photographer unknown

Big Alex McDonald, 'Klondike King,' dies in debt

Alexander McDonald, a big, lumbering and unschooled man, became known as the King of the Klondike after in short order becoming the largest landowner and biggest employer in the Gold Rush town of Dawson. He died in 1909 with assets of only thirty thousand dollars and debts far in excess of that. He had little use for gold, but treasured land. He carried to his grave title to a great deal of now worthless land, but little of the glittering mineral that for a time had made him wealthy.

Born in 1859 in Antigonish, Nova Scotia, he was the son of immigrants from Scotland. At a young age, he traveled to Colorado where he prospected for some time without success. When word came of the discovery of gold in Juneau, he went there in search of the yellow stuff but again came up empty. He then moved to the Yukon and was there when he learned of the Klondike promise.

Reaching the new settlement after most of the streams had been filed on, the big Scotsman came across a Russian who had staked Number Thirty Eldorado, twenty-nine five-hundred-foot claim lengths above Discovery at the mouth of the richest stream in the world.

Its owner had exhausted his money, his energy and his hopes in digging shafts but finding nothing of value. Tired and discouraged, he sold McDonald a half interest in the claim for a side of bacon and a sack of flour.

Slow of speech, the prospector had a habit of saying "No" to every proposition and then rubbing his chin in thought. This gave the proposer occasion to offer arguments in support of his proposal and also gave McDonald time to think. It came to the big man's advantage in this case and in many others that followed.

McDonald had no desire to personally mine on his holdings, so he hired workers or gave leases ("lays") for a percentage of what was recovered from the workings. This first effort struck pay dirt and in six weeks' time McDonald had taken in forty thousand dollars.

The mine-owner was credited with creating the lay system—the name given to the practice of leasing all or a portion of a claim for a percentage of the take. McDonald's terms called for a fifty-fifty split.

After his success with Thirty Eldorado, McDonald was approached by the owner of Six Below Eldorado, offering a lay of thirty-five feet of that promising claim for a down-payment of five thousand dollars. Using the ploy of needing time to think about it, McDonald approached Ron Crawford, a former law clerk who was setting up an office to prepare legal papers for clients. As security, McDonald offered lays on two claims and a mortgage on Thirty Eldorado. Crawford, who had recently arrived and had no available money of his own, in turn used McDonald's tactic and asked for time to think about it. Crawford then went to a neighboring saloon and borrowed the five thousand on the basis of McDonald's mortgage. He also sold part of the share on Twenty-seven Eldorado for another five thousand in gold. It turned out to be a win for all when each of them prospered from his investment in the three-way deal.

McDonald's earnings went toward buying more claims and other investments. He became a large stockholder in the Yukon Telegraph Company, profiting from the ability of the stampeders to communicate with the Outside in a matter of hours rather than months. He built a large building in downtown Dawson. A devout Catholic, he contributed the cost of building a chapel. Later, when Father William Judge proposed establishing a hospital to care for ill and injured miners and the growing population, McDonald dug heavily into his bank account to make it possible.

When fire devastated Dawson on Thanksgiving of 1898, McDonald's downtown building was among those lost. The blaze started in the upstairs of a nearby hotel when dance hall hostess Belle Mitchell threw a kerosene lamp at a rival in a fit of rage. The disastrous fire was cause for the *Klondike Nugget* to begin a crusade to establish a fire department. Dawson immediately began to rebuild. McDonald replaced his building with two others, one of which he leased to the government to be used as a post office. He again contributed a large sum to help Father Judge rebuild the church.

As spring approached, another fire swept Dawson. It, ironically, was started by the same woman responsible for the earlier one. This time, Mitchell left a burning candle sitting on a block of wood. The resulting conflagration was worse than the first one.

In the interim, a fire department had been established. The pumper was brought into service, but the crew assigned the duty of keeping open a hole in the river ice had not done so. It took considerable time to thaw and punch a new hole. Even so, after a short period of pumping water, the hoses froze.

Sam Steele of the North West Mounted Police post at Dawson took command of the situation. He ordered that two buildings in the path of the flames be blown up. The Aurora Saloon and McDonald's building were reduced to splinters in order to save the remainder of the town.

McDonald took the time to travel to Europe during the winter of 1898-99. He visited Paris and enjoyed the night life, then went to Rome where he was granted an audience with Pope Leo XIII. The Pope, having known of his generosity toward Father Judge's projects, made McDonald a Knight of the Order of Saint Gregory.

From Rome, McDonald went to London where he was intrigued by the elevators, referring to the contraptions as "heists." He was also intrigued at the sight of a pretty twenty-year-old girl named Margaret Chisholm. He married the daughter of the superintendent of the Thames Water Police in April, 1899—he being enamored of her beauty and she of his wealth.

McDonald's standing made of him a dignitary sought after for momentous occasions. An example was the occasion of two wealthy visitors to Dawson. Mrs. Mary E. Hitchcock was the widow of an admiral in the United States Navy. The other lady was Edith Van Buren, niece of the former American President. Among an unusual assortment of baggage brought with them was a tent so huge it could not be erected on the Dawson side of the river. It was raised on the opposite shore and from there the socialites reigned and entertained. Big Alex McDonald was a coveted guest at dinners, large and small, that were held there.

In telling of the first tourists to visit Dawson, author Berton reported this interesting item:

"The Salvation Army, meanwhile, had dispatched a troupe to Dawson, and these bonneted servants of the Lord, covetously eyeing the [tent's] marquee, summoned up courage to ask the ladies if they might use it for their Sunday services. The ladies were happy to oblige. The following Sabbath, as the voices were raised in prayer, it was noticed that the pigeons [part of the menagerie accompanying the visitors] had escaped from their cages and were fluttering above the heads of the uneasy congregation. One of them finally perched on the music box, which mechanically responded with "Nearer My God to Thee" and the entire assemblage rose and repeated the grand old hymn, which they had already sung."

Even though he was uneasy about speaking even in polite conversation and definitely terrified of doing so in public, McDonald was asked as the leading citizen to officiate at special occasions. When a dinner was held to bid farewell to the outgoing superintendent of the Mounted Police, Sam Steele, the organizing committee rehearsed their man thoroughly. When it came time to present the gift of gold, the honoree was told, "Here, Sam, take this poke. Good bye." Similarly prepared in advance of a presentation to the visiting wife of the Governor-General of Canada, she was told, "Take this poke. It's trash."

Although McDonald died owing more of the "trash" he disdained than he owed, his widow, who had remained in England, lived comfortably for the rest of her days. Businesswoman Belinda Mulrooney some time earlier had persuaded the King of the Klondike to purchase a life insurance policy. That investment, of all the others, proved to be the most wise—at least for his widow.

'Lucky' Anderson buys 'worthless' claim, gets rich

Charley Anderson was a thirty-seven-year-old Swedish immigrant who had mined at Fortymile for several years and decided to remain there while others joined the stampede to the Klondike, considering that a bird in the hand is better than two in the bush. His claim no longer producing as much as before, he turned to drink, imbibing more each time he heard word of big findings at the new camp upriver.

One night in 1897, he was about ready to go under the table when spotted by two men who had just arrived. Al Thayer and Winfield Oler had staked claim Twenty-nine Above on Eldorado. Sampling in the shafts they dug failed to show any promise and the two decided it was worthless. Hoping to find a sucker, they convinced the stupefied Swede to hand over eight hundred dollars and gave him a bill of sale for the claim. The next day, the hung-over miner looked up the Mounted Police constable and asked him to get his money back. The Mountie looked at the paperwork and, seeing Anderson's signature affixed to acknowledge the sale, could offer only sympathy and sage advice to drink more moderately in the future.

Wanting to investigate his unintended investment,

Anderson headed upriver and sought out the claim he now owned, two-and-a-half miles upstream from George Carmack's discovery and adjacent to the one that made Alex McDonald rich. He burned a shaft down to bedrock. What he found made him hurry to invest another eight hundred dollars for some lumber and salvaged nails from a burned building, needed to build a sluice. It turned out that the "worthless" ground had a million dollars in gold lying on bedrock. He quickly was branded as "The Lucky Swede," a nickname that stuck for the rest of his life.

The new Klondike star was flush with money, but vehemently denied the title whenever someone referred to him as a millionaire. He was said to always respond in broken English, "I neffer had a mill-yun; da mos' I ever had vas nine huner'd t'ousand."

Anderson, now sought after by every dance-hall girl in Dawson, had an eye for Grace Drummond. Gracie, a dancer and drink-hustling b-girl at Swiftwater Bill's Monte Carlo dance hall and saloon, was the girlfriend of gambler Edgar Mizner. Anderson had no respect for Mizner, the former manager of the Alaska Commercial Company trading post at Fortymile who was called "The Pope of Alaska." Mizner had angered the miners by cutting off the easy credit extended by the previous manager, Jack McQuesten, a well-liked man who freely grubstaked prospectors.

Gracie finally agreed to become Mrs. Anderson, but only after he acceded to her demand that he deposit fifty thousand dollars into her bank account. He also promised to take her on a honeymoon trip to Europe, adorn her with thousands of dollars' worth of jewels and build her a nice home. They were united by Bishop Bompas of the Anglican Church on June 6, 1899. They traveled to

Paris, London, New York and San Francisco, where he built her a twenty-thousand-dollar turreted castle on the outskirts of town. He returned to the gold fields of the Klondike, leaving Gracie behind with the keys to the safety deposit box containing a small fortune in gold, jewels, stocks and bonds.

When the Swede went back to rejoin his wife, she locked him out and filed for divorce. He argued that she had taken all his funds but she countered that his habit of dropping cold gold nuggets into the cleavage of women who caught his eye was distasteful. The "Lucky Swede" was ordered to pay her three thousand dollars per month in alimony—a term that had to be explained to him.

Anderson's luck continued to go south. His claim was worked out. In 1906, the San Francisco Earthquake cleaned him out when the real estate in which he had invested burned or was demolished. He existed thereafter on odd jobs. When he died in 1939 at age fifty-five, he had a laborer's job that paid three dollars and twenty-five cents a day. The one-time millionaire—or as he would say, nine-tenths of a millionaire—left an estate of just seventeen hundred, thirty-one dollars.

Oler, one of the two men who thought they were suckering Anderson into buying their claim, could not get over the move. He never hit paydirt and died penniless in the Pioneer Home at Sitka.

Missionaries join Rush, search for souls, not gold

The Word of God was taken to the north by soul-seeking missionaries of various persuasions long before the riches of the streams brought hordes of fortune-seekers. Even before Alaska was purchased in 1867, the Russian Orthodox Church was well established in the northwestern portion of North America. The presence of the Anglican branch of the Church of England, from whose influence the Pilgrims fled to the colonies, was strong in the Canadian expanse across the top of the continent.

After the purchase of Alaska in 1867, both Catholic and Protestant missionaries went to the northland to spread their beliefs among the people of the new possession. Some servants of the Lord stationed along the length of the mighty Yukon River transferred their bases to the goldfields where their calling was desperately needed. After gold was discovered and the stampede followed, others flocked there to minister to needs of the migrants. There were also many lay people whose good works were valuable to the life of the various camps.

The original St. Mary's Catholic Church and hospital built in 1897 by Father William Judge. The buildings were destroyed in a fire and replaced. Dawson City Museum,1996.39.24

Jesuit Father William Judge, the 'Saint of Dawson'

The work of one disciple earned for him the title "Saint of Dawson." He was a Catholic priest of the Jesuit order, fully committed to the principles set down by Saint Ignatius, its founder. He disdained material things and his life was fully dedicated to helping others in complete humility.

Father William Henry Judge was born to a strong Catholic family in Baltimore, Maryland, on April 28, 1850. He studied in many areas and learned many different secular skills that were to be put to use in his work as a clergyman. At age forty, he volunteered for service in Alaska, doing whatever and serving wherever was decreed by his superiors. After a lengthy sea voyage around the tip of South America to San Francisco and then to St. Michael, he eventually arrived at Holy Cross Mission on the Yukon River. There he joined three priests and three nuns in teaching some fifty students and ministering to residents of the community.

Two years later, Judge was sent to Nulato to build a church in that Yukon village. He travelled extensively, visiting with families in a wide area. Life was hard and conditions were difficult, but he found pleasure in his work. From Nulato he was assigned to Circle City and then to the mining camp at Fortymile. He was challenged by the assignment to the quickly-growing boom town just inside the western border of Canada.

After working with the Native populations in the earlier stations, the mostly-white miners and traders at Circle City in Alaska and Fortymile in Canada presented a lifestyle that contrasted vastly with the practices of the Natives. In both places, the population was primarily made up of Americans. A few of the people had struck it rich. The remainder were divided between workers for wages and those who were still wandering and seeking whatever was to be their fate. Among the latter, many needed care and support from the meager resources available to the priest. The kindly priest was there for them.

Word of the discovery of gold on Rabbit Creek came when the claims of Carmack were filed with authorities at Fortymile. The news quickly spread to Circle. In both places where earlier discoveries had created minor stampedes, pay was thinning as the streams showed fewer returns. In short order, prospectors headed upriver from the established settlements to the new creeks awaiting exploration. Father Judge followed his flock.

At the age of forty-five, the priest was said to have aged well beyond that number. Frail since a child, the rough life and hard work in the northland's harshness were taking their toll. Yet he always offered a pleasant countenance and a kindly greeting to everyone he met.

With little to carry in the way of personal possessions, he loaded a toboggan with supplies. With the help of one dog, he pulled the sled through snow and slush, over rocks and through swamps and across streams and tundra to the town of Dawson that was springing up forty-eight miles away.

The priest acquired three acres at the north end of the town where he began drawing up plans for a hospital, a church and a residence. Nine workmen were engaged to help with construction but the priest worked right alongside them. He found a great need for all three structures, but in reality the residence was used as much for housing the sick as the man for whom it was intended. More often than not, the priest slept on the floor in a hallway while a sick person occupied his bed.

Judge put in long days, rising at five o'clock in the morning, holding Mass at seven, then tending to the work at the hospital until eleven at night. The hospital was completed on August 20, 1897, and averaged twenty patients a day. During the winter, the number grew to fifty with cases of frostbite, injuries to miners occupied in thawing and digging shafts, falls, and fights among men with "cabin fever."

The priest was always on call, ready to respond to anyone who needed nursing or just wanted to talk. He prayed over them, offered encouragement to those who were recuperating and gave comfort to those who were dying. He was faithful to the church's calling and held regular services. He was too often called upon to officiate at funerals.

Help in tending to the sick and injured came the next summer when several Sisters of Ann came along with some committed lay ministers to help with the church. The added presence was welcome indeed when a typhoid epidemic broke out that same fall. The hospital was seeing one hundred and thirty-five patients a day during that dreadful time.

The summer of 1898 was disastrous for the overworked priest. Fire destroyed the hospital in June. With the help of Alex McDonald, Dawson's wealthiest man, and other donors, an even larger building was ready for use in just ten weeks.

As the year was drawing to a close, the people of Dawson decided to honor the man who had worked so hard and done so much so willingly and had so worn himself out. The priest's robes were tattered and patched, and the citizens decided that a new suit would be an appropriate gift that he could use. They sent a tailor to measure him but the priest declined, saying that as a Jesuit, he was prohibited from accepting gifts and would get by with what he had already been provided. The organizing committee directed the tailor to guess at the measurements and proceed with the suit. They also ordered a sealskin coat, cap and gloves to keep the aging man warm because he always dressed lightly.

The clothing was completed and presented to Judge. He persisted in declining, but finally accepted the gift reluctantly after much pleading. One argument advanced by the donors had been that "most of the contributions had come from Protestants, who wouldn't know the difference and therefore would not be offended." For a formal tribute to the priest, a minstrel show was arranged for Christmas night. Joe Cooper offered his Tivoli Theatre free of charge, even though that usually was his biggest night of the year. The priest attended the show, dressed in his new attire, ate heartily, and laughed at the jokes, even more-so at those made at his expense. The next morning he appeared at Mass in his tattered old robe, the new suit carefully tucked away.

Two weeks later, the Saint of Dawson came down with pneumonia. Death came on January 16 and Dawson mourned as it had never mourned before. The ground at the church was frozen solid and it took two and a half days to dig the grave, but there were plenty of willing hands there to accomplish the sad task. In what to him would be a major affront in life, Father Judge was buried in a casket that cost a thousand dollars, made of the finest material that could be found. It is not known if he was laid to rest in the new suit given by his adoring flock or in the tattered robe in which he felt more comfortable.

The grave of the Saint of Dawson is on the site of the old church and hospital, at the north end of Dawson. It is marked by a large stone with a plaque honoring his memory, provided by the citizens of Dawson and the government of Canada.

Rev. Samuel Hall Young opens Presbyterian church

The first Presbyterian Church (USA) congregation in Alaska was organized at Fort Wrangell by Rev. Samuel Hall Young. Born September 12, 1847, in Butler, Pennsylvania, he preferred to be known by his middle name. He was sent to the Klondike in 1879, some eight years before the discovery that started the Gold Rush. While there, he met John Muir and accompanied the naturalist on many exploratory hikes. Among other notable discoveries was their finding Glacier Bay, one of nature's marvels which continues to draw visitors today. Muir is credited with saving the missionary's life when Young fell over a cliff, dislocating both arms and landing on a narrow ledge.

After a sabbatical in the States, Young was sent back to the Klondike in 1897 where he established a church at Dawson. When it was realized that the church was in Canada, the Canadian Presbyterian Church sent missionaries to carry on the work Young had started. Heading downriver to St. Michael, Young visited camps and towns. Upon returning to the States, he urged the Home Mission Board to send several missionaries to the Yukon and surrounding area. He spent the winter spreading word of the area's needs and raised six thousand dollars for use in the mission field.

Young advocated for creation of the Yukon Presbytery to oversee the denomination's work in Alaska and it was approved in 1899. When the mission churches of Saint Lawrence Island, Point Barrow, Eagle and various villages on the northern coast met at Eagle on July 26, 1899, Young was elected as the first moderator. Upon hearing of the gold strike at Nome, he left for that camp with the intent of opening a mission there. However, under the agreement forged by Rev. Sheldon Jackson, Young was told to leave that task to Rev. L. L. Wirt, a Congregationalist.

When typhoid struck the new camp at Cape Nome, Wirt was on his way to Seattle to buy lumber for a church. An elder was holding services in his stead. Still there, Young conducted eleven funerals in one week's time, holding services in a warehouse. For some fifty days, he was the only minister in Nome to care for the sick in addition to holding Sunday services.

From Nome, Young went to Fairbanks where he organized the First Presbyterian Church of Fairbanks in the new camp established by trader E. T. Barnette.

Still standing as a Canadian national historic landmark is the shell of St. Andrew's Presbyterian Church at Lake Bennett. It was built in 1899 by Rev. J. A. Sinclair to minister to the stampeders who made their way through the passes. It was at that lake where they built vessels of every description to carry them and their goods on down to Dawson and the creeks beyond. The church was used for both Presbyterian and Anglican services. In the summer, travelers sought blessings and prayer.

Anna DeGraf's twenty-three-year-old son headed for the Yukon in 1892, saying he would return in two weeks, but was never heard from again. She went north to search for him without success, supporting herself as a seamstress.

Photographer unknown

Conscience of Yukon Anna DeGraf searched for son

Anna DeGraf was one of the first white women to climb the Chilkoot Trail, in 1892 at age fifty-three joining a small crowd of gold-seekers who were the forerunners of the great stampede to come. Her search was not for gold, but for the son who had bade her farewell in July, saying he would return and see her in two weeks. Although she was to twice hear that he was remembered as stopping at different places in the north, she was never to see or hear from him again. She died at age ninety-one, still wondering his fate.

DeGraf wrote of her quest during a quarter of a century in the northland, ending the script with the hope that he or someone who knew of him would read it and contact her. Unfortunately, the memoir was not to be published until sixty-two years after she had reunited with her son in the Hereafter. A copy of the manuscript was found in a long-neglected steamer trunk and edited by her great-great-grandson, Roger S. Brown. It was published in 1992 under the title "Pioneering on the Yukon."

The intrepid pioneer was born Anna Lotsch in Germany in 1839. At age twenty-six she joined her husband who had migrated to the United States and served his new homeland in the Civil War. Her life was to bring many heartbreaks. She lost her first child in infancy, she and her second child survived attacks of cholera, the family's financial position was wrecked in the depression of 1893, and her husband was murdered while prospecting near Yakima, Washington. She settled in Seattle where she started a dressmaking business in her home, only to see everything lost in a devastating fire. Her Teutonic spirit undaunted, she bought a new sewing machine and resumed her trade despite sustaining a severely broken leg.

Receiving no word from her adventure-seeking son after his promised two weeks had gone by many times over, DeGraf decided to go to the Klondike in search of him. Her earnest inquiries to various officials not bearing fruit, she boarded a ship bound for Juneau to follow up on a rumor that he might have been seen there. Expecting to need a source of income, she took with her the sewing machine on which she depended. She was to stay in Juneau for two years, always inquiring of itinerant miners and trappers in hopes of hearing of her son.

It is easy to see how the woman came to be known as "Mother" to nearly all who met her. Her age, the fact that she was looking for a lost son, and her kindly mother-hen approach to life all were involved.

DeGraf, who herself often needed comfort on the trail, was quick to aid others. Even those who at first were not helpful became her champions. Such was the case in one of her trips to the creeks to check with miners in hopes of finding news of her son. Unable to keep up and separated from the group with which she was traveling, she became lost among a maze of sluiceboxes lining the banks of the streams. Finally she spotted a lone figure cranking a windlass that was lifting material from a shaft. She called to him for help, only to be told in a gruff voice, "Too cold." Again pleading for him to take her to the roadhouse where she was to meet her companions, she received the same response. Taking a new tack, she asked if she could buy him supper and a warm cup of coffee when they reached the rendezvous point. That approach worked and she was guided to her destination. Once there, she upbraided her younger traveling companions for not looking out for her.

The widow's wile was often displayed. En route to Dawson by way of a boat being propelled by young men hired to assist with her trip, she worried whether she would be able to pay the duty on a ton and a half of goods she was taking into Canada. Seeing the Canadian flag flying above the North West Mounted Police post, she was inspired. Turning to one of the young men who happened to have the trombone on which he was an accomplished artist, she commanded him to play "God Save the Queen." She joined the others in singing the song so patriotic to subjects of the British monarch. The Mounties saluted and waved them to shore, then invited them in for tea. The social hour past, the party was wished a hearty "Cheerio" and departed without any mention of duty. A couple of years later, she passed the same post on Christmas Eve and was recognized as "the lady who loves our queen." She was invited to remain overnight and enjoy dinner with them the next day.

DeGraf made several trips Outside, buying material and goods to sell. Returning from one of those buying trips, she rented space and prepared to open a store. Before a dollar's worth of goods could be sold, one of the several devastating fires to hit Dawson devoured her store. Undaunted, she solicited orders for dresses and started anew.

Her customers included both the wives of Dawson's wealthy miners and the dancehall girls and actresses who spent long nights entertaining the miners. There was such a demand for dresses that she worked day and night. Friends and other concerned residents often asked if she ever slept. Her Sundays were kept open, however. She rose early, taking walks to enjoy the scenery that so greatly impressed her throughout her time in the north.

In Dawson, DeGraf met Alexander Pantages, operator of the Orpheum Theater and producer of vaudeville shows. She was contracted to make costumes for the performers and became friendly with many of them. Pantages went on to own a chain of theaters around the country and DeGraf was thrilled when she came on the Pantages Theater in Seattle on one of her buying trips. She had a long visit with him and one of the actors she had come to know in Dawson. After she left the north for good in 1917 she continued to produce costumes for his troupes. In her memoir, the gentile lady failed to mention by name the beautiful showgirl who was the constant companion of the showman, Katherine Rockwell, better known as Klondike Kate. Years after he left Dawson to found his chain of theaters, Kate sued Pantages for breach of promise. Instead of taking as his bride the woman who had taught the Greek immigrant to speak English, lavished her fortune on him, and was said to have financed his first Seattle theater purchase, Pantages married a young violinist in his orchestra.

"Mother" DeGraf was one of a few white women in the gold camps when she first went north. She loved to dance when she was able to get away from her sewing machine and enjoyed attending dances. In an area sparsely populated by those of the female gender, she received many marriage proposals. She repulsed them all, the wealthy as well as those who were flat broke.

The woman's compassion showed through in several passages of her memoir. She was devastated when a showgirl was murdered by a jealous admirer. DeGraf had been asked to make two identical costumes for this girl and for another, the two at the time preparing for a Christmas performance before a big crowd. The gowns had just been delivered when the admirer burst into the victim's room and took her life.

In another instance, the dressmaker came on one of her customers, a showgirl, lying passed out, face-down in a ditch. She fished the young woman out, took her home, changed her into dry clothes and laid her on DeGraf's bed to sleep it off. In the morning, she had a long discussion with the girl about her choices. Securing affirmation that the girl wanted to change her life, DeGraf took her to see Father Judge at the Catholic church. Father Judge heard the girl's confession, got her to sign an abstinence pledge, and they saw her off on a return trip to her home.

When the gold played out and miners departed for Nome or points south, DeGraf found employment sewing furs for the Northern Commercial Company. In 1917 she made one last visit Outside to visit her granddaughter in San Francisco and was there for the arrival of a great-great-granddaughter. She continued to sew furs and serve as wardrobe woman for Pantages until, at the age of ninety, she could no longer handle the work. She died a year later, her place in history well-earned but now virtually forgotten.

Trader John G. Healy infuriates, then aids miners

John Gerome Healy was a no-nonsense frontiersman who had traveled across the western United States for months before learning of the discovery of gold at Juneau, a spot on Southeast Alaska's Panhandle. He headed there in search of opportunities before moving on to Fortymile, a camp on the Yukon near the Alaska border. From there he moved to Taiya Inlet, north of Juneau, on the correct hunch that it would become a major factor in any coming boom. He established a trading post at the town of Dyea—a name whose spelling came from the Native tongue phonetics for that spot. In 1897 when the first stampeders climbed the Chilkoot Trail, the trading post was the only building there. Two years later, Dyea's brief boom had faded as construction began on a railroad from nearby Skagway to Dawson.

A man of many talents and little hesitation, Healy had fought marauding Indians as well as white rustlers and rowdies in Montana. In addition to being a soldier, his background included stints as hunter, trapper, prospector, whiskey trader, guide, scout, trader, and newspaper editor.

In a large area straddling the Montana-Alberta international border, he established a trading post he named

Fort Whoop-Up. The name pretty much represented what went on in the settlement controlled by the free-spirited trader. Out of the Army, he did not take kindly to authority. When an armed band of hunters tried to take over the settlement in a dispute over rights, Healy decided the matter with a simple act that did not require bloodshed. He just hinted there could be a great deal of it. When the challengers threatened gunplay, Healy calmly removed the top of a keg of gunpowder and held a lighted cigar over it. His adversaries beat a hasty retreat.

As a frontier sheriff, Healy was relentless in dealing with the bad men, causing a great deal of concern among the citizenry of the small communities.

Although the date and place of his birth are not shown, he was described as "aging" when he made his way north. That did not slow him or mellow his methods.

At Fortymile, a settlement populated primarily by Americans even though in Canadian territory, there was no law. Miner's meetings decided issues of concern to the camp. Anyone could call a meeting simply by posting a notice. Meetings were held in one or another of the roadhouses. The bar remained open during the proceedings, a fact that often contributed to the decision rendered. Historian Pierre Berton noted that as often as not, the meeting ended with the person who called it being fined. The amount generally was that which was sufficient to buy a round of drinks for the assemblage.

Healy was once hauled up before such a meeting after a complaint from a young servant woman brought from Outside to help the trader's wife. The servant liked to dance and often stayed out late, causing a curfew to be set. When the woman did not show up on time, Healy locked the door. She complained, a miner's meeting was called and the majority came out in favor of the servant. The penalty was to have Healy give her a year's salary and buy her a return ticket home.

The order did not sit well with the trader. He did not resort to holding a cigar over a keg of gunpowder, but found a way to get even that had more effect. Sam Steele, the head of the North West Mounted Police, was a friend of Healy's from his days at Whoop-Up on the border. A dispatch was sent, requesting that a detachment be sent to Fortymile to bring law and order, Canadian style. A similar request had been made to the government in Ottawa by William Bompus, a bishop of the Church of England, who complained that the miners were teaching the Natives to make liquor. The Mounties complied and miner's meetings no longer dispensed "justice."

Healy was not popular with the miners at Fortymile. He had formed a partnership with a Chicago pal to form the North American Trading and Transportation Company to compete with the Alaska Commercial Company. Until then, ACC had enjoyed a monopoly in the northwest. With competition, prices were lowered. While the miners liked being able to save on their purchases, Healy could not overcome opposition to his hard-nosed business practices. He and his managers billed for goods and services and demanded payment when due. On the other hand, Jack McQuesten, manager of the ACC post, was easy with credit, was willing to wait when hardship delayed payment and was free with information received from up and down the river.

Healy and his partner planned to establish trading posts all along the river. They also acquired steamships to transport passengers and goods. One of those boats was the *Portus B. Weare.* At the end of the first big rush to the Klondike, there was great concern that the brand-new town of Dawson was not prepared to handle the number of people who had rushed there. An early freeze-up of the Yukon had stranded those who lingered and prevented vital shipments from getting through from St. Michael.

Healy tried to comfort those who were worried that starvation would wipe out the town, insisting that enough food was available. For those who were desperate to leave, he offered cut-rate fares to be available as soon as the river opened. He ordered the *Weare* to be ready to bring food and necessary supplies upriver as soon as it was able to safely make the trip. Thankfully, his prediction that food supplies were adequate proved true. That fact did not appease his anger when the ship arrived, its cargo consisting mostly of whiskey and mining supplies and not the groceries he had asked be loaded.

The frontiersman became wealthy. A popular story of the time links Healy with "Swiftwater" Bill Gates, the man who turned talk overheard at Fortymile into a small fortune taken from the creeks of the Klondike. The two were in the crowd watching a high-stakes stud poker game at one of the Dawson emporia. Deciding to back the play of one of the gamblers, the two made side bets totaling five thousand dollars in gold. They and the card player all came up short. Only Gates was to gain some consolation from the deal: it was yet another opportunity to trumpet his image as a big spender.

Ed and Velma Lung face misfortune, hit big paydirt

As had many of the others who headed for the Klondike after hearing of the big gold strike, Edward B. Lung had recently lost his job. His was as accountant for a wholesale grocery firm in Tacoma, Washington. The son of a Baptist minister, at the age of twenty-one he left Rochester, New York, to go West as Horace Greeley had urged many years earlier. In Tacoma, he met, wooed and wed Velma Clements, a member of a well-to-do family that also was to fall on hard times.

Insight into the realities of the Gold Rush is given through two books written by their daughter, Ella Lung Martinsen. "Trail to North Star Gold" is a sequel to "Black Sand and Gold," a serial written for the *Alaska Sportsman* magazine. The former is based on Velma Lung's recollections, research, and return visits to the Yukon by the author, who was the first white child born on Dominion Creek while her father was mining there. The latter was based on Ed Lung's journal.

Lung was among the first wave of stampeders from Outside, sailing to Skagway aboard the *SS Mexico* on May 24, 1897. It would be two months yet before the *Portland* and *Excelsior* landed with more than two tons of Klondike gold aboard, generating the headlines that started the rush.

Even though he had a head start on many of the fortune-seekers who were spurred by the sensational news, Lung found that the already fabled creeks were staked for most of their lengths. He did, however, file on Number 7-82 Bench 2, Tier 8—Below Bonanza, as recorded on September 3, 1897. He sampled the ground but did not find anything exciting. He continued to “prove up” on his claim by performing some work within every five-day period as required by law, but also prospected other creeks in hope of finding something more promising. He formed a partnership with Joe Latshaw, an experienced prospector who had been in Alaska for some time and rushed to Dawson from Candle. Candle is located on the Seward Peninsula about fifty miles above Cape Nome which later would be the scene of a major find.

The pair obtained lays on several claims but found little in the way of values. Lung sold his Bonanza claim for a few hundred dollars to get him through the winter. He began building a cabin on Dominion Creek where they were working a claim on shares.

Two years and one day after he had left home, Lung wrote to his wife, asking her to join him in the Klondike. She and their young son made plans to head north in August. Although her mother cautioned against the trip, Velma could not be dissuaded from going to join her husband. Her mother had been correct in pointing out that her daughter had led a sheltered life and did not know how to cook or maintain a household. But no matter how often the earnest plea was made, it did not deter the younger woman. A family friend, Doctor William W. Misner, who had twice traveled to the Klondike, arranged passage for the two on the *Rosalie*, a small freighter departing from Seattle. Although little Clemy—whose given name was her maiden surname, Clement—was only four years old; he was a bright boy who loved to draw.

Velma recalled that the little ship twisted, pitched and swayed when a strong wind blew up soon after the *Rosalie* cleared Puget Sound. Having read every account about the place where her husband was laboring, she remembered tales of lost ships sailing to and from the northland. During the past two years, hundreds of vessels had been pressed into service to haul the passengers flocking to the far-off creeks. Many of those ships were far from seaworthy and many of their masters had no experience in northern waters. Disasters made headlines that drew the woman's eyes to the sad and scary details.

On board, the passengers were generally a friendly assortment from various walks of life. One very attractive woman, who introduced herself as Lilly Davenport, especially caught the eye of young Clemy—who, despite his mother's remonstrations, stared steadily at the well-groomed woman with long hair. His mother was struck by what she saw as she watched Lilly turn the pages of a book she was reading. Each of the woman's ten fingers was adorned by a ring, three of them having two rings for a total of thirteen of assorted sizes. All were dazzling and spoke silently of great expense. Velma was shocked to learn when they arrived in Skagway that their fellow passenger was none other than the fabled "Diamond Lil." She was appalled that the woman with all the rings was not Diamond Tooth Gertie, a popular showgirl, but Diamond Lil Davenport, a person of even worse repute, a practitioner of the "oldest profession" whose services added many nuggets to her purse and diamonds to her fingers.

In Skagway, Joe Latshaw met the new arrivals with the advice that his partner was working on the cabin in preparation for his family's arrival. He took her to "Ma" Pullen's boarding house where she was able to spend the night. She found the landlady to be quite interesting and the source of a great deal of information about the town. It was with relief that she learned that law and order had come to the port city and the dangers posed by Soapy Smith and his gang were no longer a threat.

The trio caught the train to Lake Bennett on September 1. Latshaw explained that considerable changes had been made at Lake Bennett since the first stampeders reached that point. Those beginning the water journey previously had to build their own boats. Commercial steamers now carried passengers. The newcomers and Latshaw boarded the *Nora* for the first leg of their journey. Arriving at the Whitehorse Rapids, they and the other passengers disembarked and climbed to a ridge on the cliffs rising over the river. After a portage of two and a half miles, they descended to board a sister ship, the *Flora*, for the remainder of the trip. Ed met the steamer when it arrived at Dawson and showed his wife and son around the booming town. They left the next day for Dominion Creek and the cabin of which he had written fondly.

The "dear little cabin" of which Ed had boasted was a disappointment as Velma surveyed the small, dimly lit one-room structure with its bare minimum of primitive furnishings. He explained that the cabin he was building was farther up but she would have to wait to see it for another sixty days. That cabin turned out to be a big improvement and was to be their home for the next six years.

At first, there was only one other female who lived on Dominion Creek and Velma and her little son were a rare sight on the streets of Dawson. They were comfortable at their home, however. On the Sabbath, the Lungs entertained miners from neighboring claims and these visits often turned into Bible readings and hymn sings. Velma usually gave a spiritual message and soon became known as "The Little Preacher." The meetings were later moved to the roadhouse where there was room for the growing number who wanted to participate. During services, shelves displaying the stock of spirits were discreetly covered with bedsheets. As the years passed, Clemy was able to attend school. The passing years also saw the Lung family increase in size. Joining the household was Ella, and later, in January of 1903, Paul.

The Lung-Latshaw partnership struggled as they leased various claims and sank shafts in search of the elusive paystreak. They at times hired others to work alongside them in thawing and digging shafts, then mucking out drifts along bedrock. What they were able to recover from the black sands was barely enough to meet the payroll, provide food, and buy supplies to carry over another year. The two gained a reputation as honest men who treated their workers well. There was always the hope that the next shaft would hit big pay and make them rich. Only a few of the stampeders were able to hold onto that vision and not give up in discouragement.

After the spring clean-up on the two claims the partners were working, the results were so lean that they decided to abandon them and attempt to find another location. The choices were quite limited as syndicates with large machinery or hydraulics had bought up many claims, even ones that previously had been worked over.

Their research finally came up with Number Ten Above Lower Discovery, owned by Fernand de Journel. They wondered why the claim had lain idle for some time but could learn no reason. De Journel was very gracious when they contacted him and agreed to lease the claim to them for a fifty-fifty split of the gold they produced. When the partners moved to the ground on the first of September, they found a nicely built two-room cabin, a large bunkhouse for workers and some equipment.

Lung and Latshaw hired a crew and began work in earnest. They had high hopes and their enthusiasm grew when they began to turn up good pans as they sampled the material coming from the holes. With steam points that now were in general use to thaw the permafrost, the work progressed quickly. Several shafts had reached bedrock and drifts radiated from them. Spirits soared as the sampling showed increasingly better results. The claim's lessor visited regularly and was anxious for them to get started with the clean-up. All wanted to see the results of their winter's efforts, but they were forced to wait until there was sufficient water to run the gold-bearing material through the sluices. Finally the men were able to start shoveling the sand and gravel, where quantities of gold could be seen glittering in the sunlight.

Just as they were getting started, a stranger appeared and ordered them to stop. He identified himself as a lawyer from San Francisco and the rightful owner of the claim. Neither de Journel nor the bank had legal rights there, he declared. It turned out that de Journel had a partner and each of them had mortgaged his share. The lawyer apparently had obtained both mortgages through questionable transactions. The situation was tense with the partners and the lawyer each strongly defending their positions. The lawyer even called in the Mounties and pressed charges against Lung and Latshaw. Known to be upstanding citizens, they were not arrested but Velma was instructed to take charge of the gold and keep an accurate accounting. An arbiter was appointed by the bank and the gold was turned over to him.

While the clean-up was still in progress, the partners were ordered to stop work. The arbiter advised that the bank's interest had been paid, the workers paid, and the claim now belonged to the lawyer. There was no money left for the partners who were ordered off the property. They were devastated, having put their faith into the system to bring justice. Consultation with an attorney in Dawson resulted in the advice that there was no hope for a favorable determination. Appealing to the United States Consul met with the declaration that the Canadian authorities had exclusive jurisdiction in the territory. Canadian authorities backed the San Francisco lawyer.

While Velma and the children returned to the "dear little cabin" that had been their previous home, Lung and Latshaw tried prospecting on other streams along the Yukon, but without finding anything promising. One evening over a campfire they discussed their options. There were few that offered optimism until Lung recalled that he had noticed a good showing at one end of Number Ten. There was a possibility it would extend over into Number Nine Above. That claim was owned by a French syndicate and Latshaw remembered hearing that it might be for sale.

The partners returned to Dawson and contacted the representative of the syndicate. The price, they were told, would be fifty thousand dollars. That sum was far beyond the means of the partners. Utterly dejected, they went home to discuss it more, then returned later to see if they might work out a deal. When asked how much they could raise, their calculation was at most a lowly one thousand dollars. To their surprise, the Frenchman said he would reduce the price to five thousand dollars—still four thousand dollars more than they could muster. After a little more discussion, the Frenchman said he would take one thousand and they could pay the balance from any gold they found.

Needless to say, the partners were elated. Their first shaft was sunk at the spot where Lung remembered seeing promising signs at the border with the claim the San Francisco lawyer had taken from them. In a fairytale ending, the showing that led them to seek out a lease of Number Nine had been the trailing end of an ancient channel. Their clean-up the next spring left the miners dancing and shouting wildly. The riffles in their sluice were clogged with gold.

The Lungs left the Yukon and returned to Tacoma where Ed began a successful real estate business and became a highly honored citizen of that city. Tragically, Clemy, a promising young artist, contracted polio and succumbed to that scourge at age sixteen.

Christmas Eve 1898 celebration at Pioneers Hall in Dawson attended by camp's elite. Photo by Eric A. Hegg
University of Washington Libraries, Special Collections, HEGG 3178

Dawson quickly became a bustling city, but residents had to deal with muddy streets. Photo by Eric A. Hegg

Businesses flourish as Dawson becomes frontier city

Saloons, dance halls and theaters sprang up almost overnight in the new Gold Rush towns, catering to men who were starving for entertainment. In August of 1896, the spot where the Klondike flows into the Yukon was a marshy delta. Five months later Joseph Ladue laid claim to a townsite, naming it Dawson in honor of George M. Dawson, a geologist who mapped the area nine years earlier. During that short period miners who learned of George Carmack's discovery of gold on Rabbit Creek rushed to the area and filed claims. The next spring, their cleanups were so rich that shockwaves went the length of the Yukon. The first miners took their bags of gold Outside, newspapers running headlines screaming GOLD! Since most of the promising ground was already staked, investors who were astute enough to see what was coming quickly drew up plans to mine all they could from the miners coming north, all expecting to pluck nuggets from the creeks and carry sacksful back home.

By the spring of 1898, when the first string of stampeders reached Dawson, there were buildings lining the main street, some businesses in large tents with smaller tents housing miners surrounding the whole.

The most elaborate of the downtown buildings were

the dance-hall-saloon-theaters. Built of logs or rough-hewn planks, they were of two stories. Most followed the same pattern, with a bar on one side of the front and gambling tables across the way. In the rear was a theater with a narrow stage against the back wall. The large room had benches lined up on the floor and most had private boxes along the walls. The second floor had living quarters, some of the rooms serving as a hotel.

In a letter home, miner Robert Kirk described one theater he visited soon after his arrival in 1898. It was the "Grand Opening" of the theater, the name of which he did not disclose. The evening started with the orchestra—which consisted of a piano, violin and flute—playing several popular songs as the place filled completely with men whose one-dollar ticket gained admission plus a drink and a cigar. The performance was described as a vaudeville-type variety show featuring six song-and-dance "artists," and clog dancers. The dancers drew cheers from the audience, he wrote, even though their talents were "a little thin." The performers seemed to realize their shortcomings and made the most of it by hamming it up. One girl especially was favored and the miners showered the stage with nuggets during her act. The show was followed by a wrestling match, with a boxing match closing the show. The fight, which was ruled a draw by the referee, was between Jim Slavin, an old-time fighter, and Pat Rooney, a Canadian. Kirk noted that the prize fight was the most realistic event of the program.

When the blue denim curtain came down for the final time, the master of ceremonies invited everyone to grab a drink at the bar while the benches were cleared and stacked against the walls to make room for dancing. The orchestra returned to play and a caller kept things lively.

The showgirls went around the room, inviting the miners to dance. The miner forked over one dollar for the privilege and the girl tucked away a token which she would turn in at the end of the night and collect twenty-five cents for each of the many that filled her purse. The dances were short, limited by the orchestra to perhaps two verses and repeated chorus. When the dance ended, the girls escorted their partner to the bar for a drink for each, the miner handing over his poke and the girl collecting another token as her commission. The miners never seemed to mind being hustled, appreciative of the chance to trip the light fantastic with a female companion, of which there were few to grace the mining camps.

Klondike Kate Rockwell

Klondike Kate Rockwell capitalizes on her beauty

One of the most beautiful women to grace the Klondike during the big rush was very conscious of the fact that she attracted the attention of men—and used that attraction to make herself rich.

Born Kathleen Eloise Rockwell in Oswego, Kansas, in 1873, she was an independent spirit from the time she could toddle around the house. Her parents, John, a railroad telegrapher, and Martha, a waitress, were both divorced from earlier marriages. They lived for a time in North Dakota, and then in Washington State. When their daughter was but a small child, the couple called it quits and Martha married her lawyer, Allison Bettis, taking Kathleen with her. Bettis was several years older than his new bride and was to become a legislator and a judge. They had a large home and Martha and her daughter experienced a lifestyle that had until then been completely foreign. Bettis cherished the little girl he called "Kitty" and gave her everything she wanted. She learned to ride and was considered a tomboy as she was growing up.

Even at a young age, Kitty showed an interest in boys. Her flirtations were serious enough to cause her mother to send her off to a Catholic school in Kansas. The spoiled child did not respond as had been hoped, but rebelled at the strict rules imposed by the nuns. From Kansas, the girl was sent to another school in Minnesota where her rebelliousness continued; expelled from that school, she was shipped off to one in California. By that time, Bettis had lost all his holdings after being caught up in the depression that was especially felt in the West. Martha bade him farewell and tried running a boarding house, but soon sold it. She took her daughter on a trip to South America to visit Kate's half-brother, the issue from Martha's first marriage. The young girl was thrilled with the voyage—especially with the attention given her by the sailors from whom her batted eyes earned many favors and special attention.

Still hoping to protect her daughter, Martha registered her in another Catholic-run school in Chile. With the child safely ensconced, the mother left for Europe, apparently in search of husband number four. Martha hurried back to the States, however, when she learned that Kitty had managed to become engaged to a young attache' at the Spanish consulate. She ordered the now sixteen-year-old girl to join her in New York. When they reunited, Kate was devastated to learn that her mother was now impoverished. Realizing that she would have to help provide for their keep, Kate turned to the classified section of the newspaper. Catching her eye was an ad that solicited chorus girls, specifying that no experience was needed. That was considered by the girl to be the perfect career. She was hired to be the page girl in a Coney Island dance hall. The comely young red-head with long black lashes and great figure soon was promoted to the chorus line. There she caught the eye of a vaudeville producer who hired her to appear in a "legitimate" vaudeville show.

For reasons unknown, Martha and her daughter, who now went by Kate, left the Big Apple and returned to Spokane where the daughter found work in a less-than-legitimate vaudeville theater. There she learned that her good looks could be converted to cash by being willing to chat with lonely customers after the show, enjoying a glass of lemonade while the customer sipped champagne. The entertainer was given a commission for each round paid for by the starry-eyed customer.

Hearing that better opportunities awaited across the border, Kate left her mother in Seattle and went across the sixty-two mile stretch of water to Victoria, British Columbia, where she had been offered a slot at the Savoy Theater. Victoria was a booming town as a result of the Cassiar gold rush that started in 1850. Its economy was bolstered, too, by being the major North American port in the opium trade, where importation and distribution of the narcotic was still legal. Hearing of the big Klondike strike, Kate joined other girls with hearts set on finding good fortune. Arriving at Lake Bennett late in the fall, they were stopped by Mounties who declared the trip downriver too dangerous for women. Not to be deterred, the headstrong young dancer put on men's clothing and, just as a boat departing for Dawson cast off, jumped aboard. The occupants of the craft were in no mood to turn back the new, non-fare-paying passenger who later described the trip as "the most exciting of my life."

At Whitehorse, the young woman found a letter waiting from the organizer of a large troupe planning a tour to Dawson and offering her a role. She turned around and went back to Victoria to join the group which offered a booking at a theater in the new boom town. During her brief but profitable travels north, Kate had already become familiar with audiences comprised mostly of lonely miners hungry for entertainment and the sight of a female. When Kate came on stage, the latter was even more pleasurable for them than the former. She knew what they wanted and how to get their hormones to kick in to the point of handing over their gold in hopes of getting more than entertainment. The old show-business adage that an entertainer should always leave them wanting more had been firmly planted in her psyche and she practiced that to the fullest extent.

Her flair for the unusual in her dances caught the attention of one "Arizona Charley" Meadows, who had just opened the Palace Grand Theater. He offered her a much better salary and star billing and she promptly moved into the star's suite upstairs over the showroom. There she developed her signature "Flame Dance," a highly sensuous exhibition in which she unwrapped herself from what was claimed to be two hundred yards of chiffon. Wearing a crown fashioned by an eager miner from tin cans with many lighted candles stuck in its spires, she whirled around the stage at a pace furious enough to keep the filmy band of cloth floating in the air until she fled the stage at the last strain of the music. She was later to write that she spent an hour each morning in the gymnasium, building up her strength to meet the demands of the performance.

In addition to the attractiveness of her figure and willingness to let her legs, shoulders and cleavage be viewed by the miners, her pretty face portrayed an innocence that belied her actions. Her voice was husky and she belted out ballads in a sultry manner. On the street and off stage, she was friendly and always encouraging in her manner. That was entirely cultivated and by design.

An interesting description of Kate and the way her fortune increased was given by miner Ed Lung. The son of a Methodist minister, he had prospected along the Yukon for several years prior to the discovery on Rabbit/Bonanza Creek. Working a promising claim on Dominion Creek, he sent for the wife and young son he had left behind in Tacoma. An off-hand remark to Mrs. Lung about a ride given to him and his partner on a stage carrying six show-girls required a confession. Lung and his partner, Joe Latshaw, had been walking back to Dominion Creek from Dawson when the women offered a ride. Both men possessing good voices, they joined in singing familiar songs. Lung sat next to Kate Rockwell while his partner sat next to Cad Wilson, another top-billed and beautiful showgirl. Not long after, Lung encountered Kate and, apparently wanting to make an impression, pulled out a small poke and poured an assortment of nuggets into the palm of his hand. He offered to give her one. Lung told his wife that she fingered each one admiringly, then chose the most valuable one of the bunch. With hindsight, he realized the showgirl had been silently weighing each of the yellow beauties as she fondled them. Even so, he described the showgirl in kind words, saying she was very young, maybe nineteen or twenty, and very pretty and very sweet. Had he known her birthdate, he would have been surprised that she was four years older than he guessed. It turned out that she gave her birth year as later and later as the years progressed.

At the Palace Grand, Rockwell met and fell head over heels with a Greek immigrant who spoke no English when he first arrived in the Klondike. Born Percales Pantages, he gave his first name as Alexander, thinking of the explorer and conqueror Alexander the Greek. Pantages was working as a swamper in the theater, filling in as a bartender when needed. He had some experience in theaters, having started as an usher and working up through the various ranks while gaining some production experience. It was common ground and a subject from which Kate could learn. They went for sleigh rides to watch the stars during clear winter evenings and hiked and picked flowers during the summer, very much the picture of two people in love. Pantages gained stature with Meadows, the theater's owner, when he offered to fill in for a local fighter who backed out of a scheduled bout with a visiting pugilist. Relieved that he would not have to cancel the show, Meadows agreed. As it turned out, Pantages had done some boxing prior to coming north. He won the exhibition with the storied visitor, becoming thereafter an often-billed attraction.

Pantages later told Kate he wanted to go Outside and purchase a theater. She agreed to back him in becoming the owner of a small showplace in Seattle and he left with a supply of her gold in hand. It was a sad parting, as the two had often spoken of marriage but never took the step. It is believed that a son came from that liaison. Kate left Dawson for a period and returned with a baby she said was the child of a woman she had befriended and who had died in childbirth. She later said she had found a good home for the boy. When Kate eventually went Outside for good, she gave her forwarding address as in care of "Lotus Rockwell in care of the Alcazar Theater in Seattle." In his many letters to Kate after Pantages left, he referred to himself as "Papa," lending more substance to the belief that the baby was theirs.

Pantages eventually became the owner of more than twenty theaters around the country.

Kate was devastated, several years after Pantages left her in Dawson, when she learned that her intended had married a young violinist in the orchestra in one of his theaters in Seattle. It was even worse when Kate learned that she had worked with the violinist in the past. That apparently was the only time in any of her many fabled engagements that the beauty had been the one who was jilted. In the wake of that rejection, Kate turned to alcohol. Before long, most of her fortune had been dissipated. She sued Pantages, not for fraud in taking her money but for breach of promise, asking for twenty-five thousand dollars in compensation. She reportedly settled for five thousand. Later, desperate enough to ask her former lover for money, he gave her the paltry sum of six dollars.

No longer able to attract worshipping fans as she had in the Klondike, she toured the country in a song-and-dance act, then joined for a while in a roller-skating act until suffering an injury. She bought a three-hundred-sixty-acre homestead near Bend, Oregon, and lived in a one-room cabin there while she made improvements. At the age of thirty-nine, Kate married Floyd Warner, a tall, handsome cowboy who was thirteen years younger. She remained flirtatious and cheated on him, bringing him to sue for divorce.

In the fall of 1929, Pantages was convicted of raping a seventeen-year-old who had come to his office seeking employment. He was sentenced to fifty years at San Quentin. While in prison, Pantages' wife, Lois, was convicted of manslaughter while driving drunk. For a reason not made clear, Kate was subpoenaed as a witness against her but was not called to the stand. Lois Pantages was convicted and sentenced to prison. Within two years, the sentences of both Pantages were overturned.

The fact that the famed Klondike Kate was expected to appear in the manslaughter trial was widely reported in newspapers around the country. That the former Queen of the Klondike was in dire straits brought proposals from two men. One was a shy miner who had, as had hundreds of others, fallen for her but had never made his intentions known. Johnny Matson, then about seventy while Kate was fifty-seven, was still in good condition and working his claims in summer and trapping in the winter, proposed by mail. Kate accepted and went to Dawson where they were married. Matson insisted that she spend the winters Outside rather than endure the cold temperatures of the northland. She spent the summers with him and lived in California in the winter. The publicity surrounding the legendary woman caused Columbia Studios to buy the movie rights to her life story. She also earned some money for coaching young starlets in come-on techniques, a subject in which she was well-versed. Matson perished in a Klondike blizzard in 1946, reportedly while heading out to mail a letter to the woman he so deeply loved. His body was found by the trapper Kate contacted when she did not hear from him.

Two years later, Kate Rockwell Warner Matson and William L. Van Duren applied for a marriage license in Vancouver, Washington. She had paid for cataract surgery for Van Duren several years earlier when he lost his accounting job because of failing eyesight. The man's wife left him when he was unable to work and could not afford the vision-restoring surgery.

Kate was honored in 1931 at a reunion of Alaska-Yukon Pioneers at their annual Stampede in Portland, Oregon She attended wearing the same fifteen-hundred-dollar gown made by Worth in Paris that she had worn in her Palace Grand debut in Dawson so many years earlier in an effort to please Pantages. When the master of ceremonies introduced her "not as Aunt Kate of Oregon, but our Kate, the Belle of the Yukon, Klondike Kate, the sweetheart of the sourdoughs," one thousand of those sourdoughs rose to their feet and applauded loudly and long. Kate died at Sweet Home, Oregon, February 21, 1957, at age eighty-four—or, as her obituary read, eighty.

Kathleen Rockwell, known fondly as "Klondike Kate," was one of the most popular showgirls in Dawson.

Eric A. Hegg photograph, courtesy Library of Congress

Gold Rush Showpeople highly admired, highly paid

Showgirls who made their way to the theaters that sprang up in the Gold Rush towns of Skagway, Dawson, Grand Forks and, later, Nome, displayed extremes of talent, voice and looks. Their nights were long and their duties were varied. Long also were their earnings. Performers in that day averaged one hundred fifty dollars per week, musicians twenty dollars per night. That compared to fifteen dollars a day paid to miners who thawed ground, dug shafts and shoveled the material from which gold might be extracted.

In addition to performing in shows, the girls were also required after the curtain fell to dance with the miners, then hustle drinks. For some of them, extracurricular activities provided even more gold dust to salt away in their socks. As to open practitioners of the shady method of earning money, there was a definite distinction between showgirls and prostitutes. While the showgirls lived in hotels or in private homes, the women of ill repute were relegated to small cabins in the red-light district. Their work was illegal, but tolerated by the Canadian authorities as a necessary evil.

Little is known of the prior history of most of the women or what became of them when they left the Yukon. It is likely that they wished to remain anonymous, desiring to hide their activities from loved ones back home or from people they might come to know in the future. Many of the legitimate actresses were known by stage names. Those of a lower class were known by often crude nicknames assigned by the miners.

'Diamond Tooth' Gertie featured in modern shows

One of the best-known showgirls to hit the Klondike went by the name of Gertrude Lovejoy, popularly known as "Diamond Tooth Gertie" because of the glittering chunk of hard mineral pressed between her front teeth. Nothing is found in her past to tell who her parents were, where or when she was born. The reason her past is not traced quite likely is because her real name was abandoned in favor of the more suggestive quality of her stage name.

There was plenty written about her presence in the northland, however. Her time there is celebrated today in Dawson where Diamond Tooth Gertie's dance hall is the setting for shows held each summer for hordes of tourists who venture to the fabled site.

A headliner at Arizona Charley Meadows' Opera House and Palace Grand Theater, Gertie kept the miners laughing and whistling cat-calls with her off-color ad libs and unabashed invitations to toss nuggets at her feet. A comely wench who showed her figure to the best effect, the men were anxious to share a drink with her after the show. Those willing to occupy one of the curtained balcony boxes—where the most expensive wines were served—were the most often to win her favor.

While there was always an enamored following, Gertie often could be seen in the company of Meadows or the United States Consul, portly and fun-loving James McCook. Both men were married, but liked being seen with the dance-hall queen. Although many marriage proposals were offered by miners, including some of the wealthiest men in town, Gertie chose C. W. C. Tabor, Dawson's leading attorney, as her husband. His income was estimated at two thousand dollars a month, but that was a drop in the bucket compared to what came Gertie's way. She lost that much in clothes and jewels when she narrowly escaped a hotel fire in 1900. The diamond in her mouth, though, was intact and she often quipped that it stood ready in case she ever needed a stake.

Despite becoming the wife of the town's most respected attorney, Gertie was snubbed by the matrons in Dawson's polite society. When the city's population began to dwindle, Gertie tired of the winters and moved to a ritzy section of San Francisco. When her husband died in a fire at the Yukon Hotel in 1917, Gertie sent only a spray of hyacinths even though she could afford to buy out the flower shop.

Cad Wilson wows miners, picks 'em clean

Another popular showgirl was Cad Wilson. A bit more is known of her past because she used that name in various theaters prior to heading to the Yukon. Billings can be found for her as early as 1893 when she did outlandish dances and a drunk act at San Francisco's Orpheum. She also was listed as playing at the Trocadero in New York City and in Victoria, British Columbia.

Wilson turned up in Dawson in September, 1898, where she appeared at the Tivoli and was an immediate hit. She was said to be less than beautiful, had a so-so figure and only a passingly pleasing voice. Any shortage of physical attributes were more than offset, however, by her coquettish appearance, her suggestive lyrics, her come-on moves and her brazen teasing of the mostly male audience. That she was out to take the miners for all she could get was made plain by her manager, comedian Eddie Dolan, who introduced her each night.

Dolan waved a piece of paper he said was a letter to the actress from her mother. The letter, the audience was told, advised her to "pick nice clean friends" in that far-off land. "I'll leave it to you boys," Dolan said with a wink and a nod, "if she can't pick 'em nice and clean." The line always drew a loud guffaw as the miners anxiously awaited the appearance of the headliner. She adopted as her theme song one with the title "Such a Nice Girl, Too." It was sung as she posed and pranced suggestively. She let it be known that she liked gold nuggets and when they were thrown onto the stage she pouted and made it known that she was looking for larger ones. The finale of her program was a throaty rendition of "There'll be a Hot Time in the Old Town Tonight" that brought a hail of nuggets. She had a practiced eye and went into a public pout if the take was less than five hundred dollars. It seldom fell short. After the show she retired to the balcony box from which the most enthusiastic shower of nuggets had come.

Wilson only had one bad review from the adoring press. When invited soon after her arrival to appear at a benefit being staged by the Elks Club, she failed to tone down her act to a level more appropriate to the occasion. The *Klondike Nugget* lambasted her performance, saying it might have had the boys in the back rows howling, but the one hundred or so ladies in the front rows were aghast and had hung their heads, the gentlemen who had brought them to the affair wishing they had not.

Undaunted, the actress continued to wow her audiences and the theater packed them in every night. And the nuggets kept coming from the starry eyed miners enticed each night to toss ones even larger than the one just picked up. The wealthy miners known as Eldorado Kings got together and had a large belt of nuggets made up for her. She wore it on stage and hiked her dress high enough to disclose a smaller nugget belt around her garter. The gift gilt belt was even put on display in a glass box in the window of the jeweler who manufactured it.

One definition of coquette is "a woman who endeavors without sincere affection to gain the attention and admiration of men." Illustrating how well that definition fit Cad Wilson is the widely reported incident in which she took the man known as "The Sawdust King" to the cleaners. The man who changed the sawdust on the saloon floors each morning frequently gambled away his earnings. One night he came away a winner, pocketing eighteen hundred dollars. He headed straight for the Tivoli, boasting to the actress of his luck. She took him to one of the boxes where the rich men sat and they drank champagne all night. She collected four dollars commission on each bottle consumed, her glass, of course, filled from a bottle containing a non-alcoholic beverage. The gold-digger sat on his lap, stroking his unruly hair and whispering sweet nothings, until his stack had dwindled to only sixty dollars. Even though he wanted to continue drinking until the last dollar was spent, Wilson had him escorted out. Whether she was just plain tired or had a brief spark of charity and wanted to leave him with enough to buy a meal is not known.

Wilson spent only one season in Dawson and is said to have left with twenty-six thousand dollars. She continued to appear in shows Outside and in 1899 went to Nome in hopes of repeating her Klondike success. Nome was different, however, with the miners too busy working or fighting off claim-jumpers to be entertained in the manner to which she was accustomed. She left the northland and disappeared into obscurity, although in 1908 her name surfaced, along with those of ten other women, as respondents in a divorce case against a San Francisco businessman.

'Diamond Lil' Davenport died alone, broke

The woman who on first sight impressed a young passenger as a genteel lady with rings on every finger turned out to be not a showgirl but the proprietor of a Skagway bawdy house. The bejeweled woman and Velma Lung were headed north in the fall of 1899 aboard the *SS Rosalie*. The woman's description was related to Ella Lung Martinsen by her mother and was quoted in Martinsen's book, "Trail to the North Star Gold."

Lung's young son Clement could not keep from staring at the woman they called "Lady Lily" even though she had not spoken to or acknowledged them. She was described as "a young woman of unusual beauty with the bearing of a queen. She was tall and stately, beautifully gowned, fair skinned, with velvety eyes and complexion like the inside of a petal of a lovely white lily."

One can imagine the chagrin of a real lady who later learned the truth about the woman known as "Diamond Lil" Davenport. Lil was almost six feet tall, buxom, and flashed a smile showing a diamond stuck in her front teeth. Her place of business was lavish and richly appointed, a talented pianist entertaining waiting clients. She catered to the well-heeled, leaving the wage-earners to the girls in the cribs on the outskirts of town.

Born Honora Ornstein to a prominent cattle rancher in Butte, Montana, Lil reportedly fled to Chicago where she fell in with shady characters. Upon learning that the new gateway to the Klondike at Skagway was booming, Lil gathered up some attractive "actresses" and headed north the year before she and Mrs. Lung made the trip. Lil reportedly had cleared thirty thousand dollars in the first season, enabling her to buy the gowns and jewels she displayed.

Lil had many beaus among those who sought her favors, reportedly changing her will to benefit the most recent of those whose fortune had been lavished on her account. She liked to present them with diamonds as a reminder that they were from her.

When the rush died down and the stream of prospectors and miners bringing out their gold fell to a trickle, Lil left for the Outside. An item in the July 16 edition of the *Fairbanks Daily Times* reported that the well-heeled madam was fined twenty-five dollars by the Juneau city magistrate "for using obscene language." She had the habit of hiring two taxis—one in which to ride, the other following along in case the first one broke down. She had two marriages, both of which failed. Her fortune soon faded, as did her face and her figure. She found employment that still kept her in proximity to money: her last job was as a scrubwoman in a Seattle bank. In 1935 she was admitted to Western State Hospital, her body suffering the effects of venereal disease and her mind wracked from the narcotics she ingested. Her diamonds were sold to pay off her medical bills. She died alone and broke in a Yakima, Washington, nursing home in 1979.

'Arizona Charlie' Meadows opens Grand Opera House

Tall, lean, mustached, with long hair and looking the part of the Wild West character he was, Charles Henson Meadows was to see the best of his many get-rich-quick schemes reach fruition in Dawson. He was the builder of the Grand Opera House, the finest showplace in the "Paris of the North."

Born March 10, 1860, in Visalia, Tulare Country, California, into a rancher's family, he learned to ride, shoot and rope, taking part in "cowboy tournaments" that were forerunners of modern rodeos. Away from home in 1881, he returned to the ranch to find that his father and brother had been killed by a band of Apaches that had raided ranches in the area. He joined traveling Wild West shows, including that of William F. "Wild Bill" Cody and in 1893 formed his own group. The show, with local Indians and cowboys, opened at the Territorial Fairgrounds in Phoenix, then went on the road. The show folded in Sacramento some months later. Meadows then went to Australia to run a Wild West show that was performing in conjunction with a circus. The circus disbanded, but the Meadows show completed the tour.

After hearing of the Klondike gold strike, Meadows rounded up a group of his cowboy buddies to drive two hundred head of wild burros north with the intent of using them to haul freight. They sailed from San Francisco to Skagway, then headed through White Pass to Dawson. Some reports indicate they ran into difficulties and lost some of the animals and their grub. How many animals and men from the party actually reached Dawson is not recorded, but "Arizona Charlie" Meadows definitely made it through.

He hired an architect from San Francisco to design an opera house that would be the finest in the Yukon. Dawson was already becoming known as the "Paris of the North" and Meadows wanted a place that fit that description. The Grand Opera House was built in only six months' time. Always on the lookout for talent, Meadows attended the opening of the Savoy Theater Group's vaudeville show in an existing opera house. The group, based in Victoria, British Columbia, consisted of more than a hundred and fifty entertainers and musicians with costumes and props. He was impressed with a stunning dance act performed by a pretty young woman who had captured the attention of the miners. A good crowd had been drawn by the advance billing for the show. The ad read, "Entertainment Consisting of Amazing and Edifying Offerings in Thrilling Scenes from Our Astounding Repertoire, Interpreted by Matchless Artists of Our Internationally Acclaimed Company. Thrills and Fun for All. Elevating and Entertaining. Gentlemen, Please Be Seated!"

Meadows also scheduled boxing matches that were popular with the miners. On one occasion, a touring pugilist was scheduled to take on a local tough. When the night arrived, however, the local man was nowhere to be found. The day was saved when a Greek who barely spoke English, hired as a swamper and sometimes bartender, volunteered to take on the visitor. To the pleasant surprise of the theater owner, Alexander Pantages defeated the professional. It turned out that Pantages had boxed in his homeland. He also had experience in staging and managing theaters—something that Meadows was quick to lean on.

Another fortunate thing for Meadows was that Pantages and the dancer enticed to leave the Savoy group and switch to the Grand had a thing going on between them. It was a strange match, because Pantages was not wealthy and therefore not one of those to whom the girl generally gave the come-hither look. But the dancer, soon to be known at Klondike Kate, the Belle of the Yukon, and also Meadows were to gain from Pantages' knowledge of show business.

It was not too long, however, before Dawson's population began to dwindle. A new strike had been made near the other end of the Yukon, at Cape Nome. The richest claims were being cleaned of the valuable mineral and were being bought up by syndicates and soon to be worked by huge dredges. Miners who struck it rich were heading to homes Outside to enjoy the fruits of their labors. Those whose dreams had not been fulfilled pulled up stakes and joined the new stampede. Non-miners—the hangers-on and service workers—no longer could afford to stay on now that gold was not tossed about as easily as in the town's heyday. Pantages left to become the owner of a theater chain—largely financed by Kate's purse. Kate was to remain for a while but was soon to return to the States, hoping to marry her lover only to be jilted. Meadows sold the opera house, now known as the Grand Palace, for seventeen thousand dollars, about a third of what it had cost to build. Even so, he was a wealthy man.

Meadows promoted various other enterprises, but none was as successful as had been the opera house he built in the Yukon. He died in Yuma, Arizona, on December 9, 1932. His demise came from infections resulting from his attempt to remedy his varicose veins by slicing away at them with his pocket knife.

More than twenty years after Arizona Charlie Meadows was laid to rest, the Yukon parliament spent three-quarters of a million dollars to restore the Palace Grand to its former splendor. It continues to be a historic shrine and tourist attraction where Gold Rush shows are put on in the summer for the benefit of visitors.

Arizona Charlie Meadows, foreground with thumbs hooked in belt, opened his Grand Opera House to a full house. He had an architect design a showplace to match Dawson's title of "Paris of the North." It cost $40,000 in 1898 but brought only $17,000 when sold as the Klondike Rush waned. Photographer unknown

'Swiftwater Bill' Gates flamboyant Klondike figure

A small man with a large ego and a penchant for pretty young lasses—and who made and spent fortunes in the Klondike and Alaska—"Swiftwater" Bill Gates was one of the most fabled characters of the times. He was shrewd in getting control of mining claims, hired other people to work them, then tossed around the proceeds in reckless abandon. "Abandon" is a word that sadly came into play more than once in his questionable activities.

The Swiftwater nickname arose from his boasts about being an expert boatman steering his way through the rapids of the Couer d'Alene River in Idaho. Little is known about him prior to his bursting onto the pages of Yukon River history in 1897. If one were to guess, he was likely in his early thirties when he appeared at Circle City the year before.

Circle City, located where the Yukon makes its northernmost loop at the Arctic Circle, was the site of a boom town that sprang up following a discovery of gold in the vicinity. Joe Ladue set up a trading post there shortly thereafter. By 1896 the settlement had twelve hundred residents. In addition to Ladue's trading post there was a music hall, two theaters, eight dance halls, twenty-eight saloons, a school, a hospital, and a library.

There is no known record of Gates' arrival or what endeavors had occupied him in the fourteen years between the Couer d'Alene gold rush and Gates' entry into the Klondike.

What is known is that Gates was destitute in the winter of 1896 and took a job as a "swamper" in the Fortymile saloon, where Bill McPhee was tending bar. It was a lowly job, one most men took on only in desperation. The duties involved mopping the floors, liming the outhouse, washing dishes, cutting firewood and keeping the stove going—and cleaning the spittoons.

Gates was short, about five feet, five inches tall, paunchy, with a round face and scraggly beard. His appearance was unimpressive and he drew little attention.

When the first word of the Klondike discovery reached the Fortymile trading post it did not create a lot of excitement. There had been other discoveries, including the one that led to the camp where they were, and none had rivalled the California rush of 1849. Nevertheless, Gates paid close attention to the stories. When George Carmack brought out his shotgun shell full of nuggets, Gates abandoned his job, found a willing companion, and slipped off into the night, heading upriver to that place in Canada with the then-unfamiliar name.

With no money but self-confidence that knew no bounds, he and six partners took a "lay" on Number Thirteen Eldorado. A lay is what miners called a lease on all or a portion of a claim. Thirteen was available at a bargain because the superstitious stampeders considered the number unlucky.

While his partners thawed ground and dug shafts to reach bedrock, Gates was already transforming himself from the destitute swamper into a wealthy man. He borrowed money at ten percent per month interest, with his lay on Thirteen as collateral. He appeared in a Prince Albert coat, top hat and white shirt with a starched collar—the only starched collar in Dawson. He gambled with abandon, tossed his poke on the bar and bought rounds for the house, and dreamed of his future as a millionaire.

There was cause for the dream. When their seventh shaft reached bedrock, the partners hit pay dirt. Abandoning the code of the Yukon to share information, they burned the wooden cribs supporting the windlass at the top of the shaft and made it known that Thirteen was indeed an unlucky number. After all that hard work, they claimed the claim panned out at only ten cents. With that, H. B. Hollingshead, the owner of the claim, who had not done much exploration work on his portion, was happy enough to sell out for forty-five thousand dollars. Gates and his partners borrowed the money. After breakup, when they could sluice the material from their dump, they repaid the loan in six weeks.

Swiftwater took other lays, hiring workers. He didn't pay them in cash, instead signing over fractions of his leases. As hard as he tried to throw money away, it kept coming in faster than he could get rid of it.

Women had never paid attention to Bill Gates. Now that he was flush with gold, he sought to remedy that. He especially liked the young, pretty ones, ones such as those showing up in the dance halls. To impress them, he loaded up carriages and took them to the mine, letting them pan out and keep all the gold they wanted.

Gates found a man who appeared to be a shrewd businessman and talked him into going into a partnership to build what would be the grandest pleasure palace in the north. It would be in the heart of Dawson and have a theater, dance hall, saloon, and gambling parlor. While the two-story building was under construction, the Monte Carlo Dance Hall and Saloon opened in a tent.

Motivating the man in the long coat, starched collar and top hat was a plan to bring back the love of his life, Miss Gussie Lamore, at 19 the oldest of three sisters who had brought their beauty and their talents to entertain the lonely miners from the creeks.

A widely-reported story about their one-sided romance is that Swiftwater learned that Miss Gussie was fond of eggs. There are many versions of the story and even the most conscientious writers who researched them failed to establish the truth. The essential is that Gates bought up all the eggs in Dawson, be it the restaurant's stock, one case or two, and either fed them to her or had them fried up so that he could toss them to hungry dogs outside the restaurant door.

The eggs apparently were not enough to win the lissome Miss Lamore's heart, leading the smitten mine-owner to offer to buy her for her weight in gold. That offer at first was spurned.

At season's end, however, Gussie finally agreed to meet Gates in San Francisco where they would be married. What the not-so-swift Bill did not know was that the love of his life was already married, having tied the knot some three years before, when she was sixteen. Gates went there planning to also buy fixtures for the Monte Carlo, but returned to Dawson only with the purchases, and not the woman who had captured his heart.

He tried his hand with both of Gussie's two younger sisters but got nowhere with them.

Swiftwater's heart did not stay broken for long. In Seattle he met the widow Mrs. Iola Beebe and her two daughters, Bera who was fifteen and Blanche who was four years older. Both, it turned out, had been in a convent and only recently left that protected environment. By the time the flamboyant "Prince of the Yukon" was ready to set sail, he had convinced both girls to leave Mommy behind and go with him. Mrs. Beebe found out and caught up with the *Humboldt* just as the ship was ready to cast off with her daughters aboard. The scoundrel who was planning to abscond with her daughters reportedly was found hiding under a lifeboat.

The widow did not go to the local authorities, but decided instead to accompany Gates and her daughters to Skagway. The women took a room at a hotel. Mrs. Beebe woke the next morning to find only her older daughter present; fifteen-year-old Bera had eloped with her Prince. The couple were married after reaching Dawson, a fact that later was to escape Swiftwater's recollection.

Gates had other things on his mind. His luck was running out. At last he was spending money faster than he could rake it in. By the end of the year, he was a hundred thousand dollars in debt and bankrupt—with a wife and four-week-old daughter as dependents. Leaving the daughter with grandma, Gates and Bera left for St. Michael, the Alaska port at the mouth of the Yukon.

In Nome, Gates managed to get a lay on a Dexter Creek claim. It appeared that his luck was returning. He made four thousand dollars in short order. Confident that the luck would transfer to the poker tables, he threw in his poke and watched its contents vanish.

It turned out that at the time, Gate's sister was in Nome with her seventeen-year-old step-daughter, Kitty Brandon. Forgetting that his wife and the mother of their daughter was waiting, Gates sought a new life with a new wife, taking Kitty, his niece, Outside with him. They were married in Chehalis, Washington.

Once more Iola Beebe chased down the scoundrel, having him arrested on charges of bigamy. It is not clear exactly how, but the charges were dropped. Cleared of charges and arranging to get divorces from both wives, Swiftwater was a free man. Iola, his former mother-in-law, wanted him to finance construction of a hotel for her. Instead, he talked her into pawning her jewelry. He used the proceeds to go back to Alaska, this time to the new gold strike at Fairbanks. He took a lay on Number Six on Cleary Creek where he took out seventy-five thousand dollars.

Swiftwater Bill was involved in several more escapades, most of the details of which are murky and forgotten. The Prince of the Yukon died in Peru in 1935 while there to pursue a fortune in silver.

Looking upstream from No. 13 El Dorado claim leased by "Swiftwater" Bill Gates and partners. Photo by Eric A. Hegg
University of Washington Libraries, Special Collections, HEG 13B

Jefferson Randolph Smith ran Skagway underworld

One of America's most notorious confidence men met a violent end on a Skagway wharf on July 8, 1898. The first man to reach the dying Jefferson Randolph "Soapy" Smith was Presbyterian minister J. A. Sinclair. Sinclair, who had only recently arrived at the Klondike gateway port, upon recognizing the moribund kingpin of Skagway crime was heard to say, "Thank God."

The man who ruled the gateway to the Klondike was born in Newman, Georgia, on November 2, 1860, the son of a well-to-do lawyer. They lived on what had been a large plantation owned by Soapy's great-grandfather. Their land and wealth went with the wind in the aftermath of the War Between the States. The family in 1876 moved to Round Rock, Texas. It was there where Jeff, then eighteen years old, was with his cousin Edwin on July 19, 1868, when the outlaw Sam Bass was fatally wounded. Bass, a train robber by trade, and his gang were scouting a bank they planned to hold up. When Bass and another robber were recognized by Deputy Sheriff A. W. Grimes and ordered to surrender their firearms, the deputy was shot and killed. Two Texas Rangers cornered Bass. He was shot first by Ranger George Herold and then by Ranger Sergeant Richard Ware. The

Smiths were standing next to Ware when he fired. Witnesses said Jeff exclaimed, "I think you got him." Indeed, the wounded Bass was found lying in a pasture and died the following day.

The man whose criminal career ended that drizzly evening in Skagway's heyday was not deterred by having seen the outlaw Bass gunned down nineteen years earlier. He drifted to Denver, Colorado, where he fell in with a slick card sharp who made of the young man first a shill, then a student in the fine art of sleight of hand. By the time they parted company, Smith was an expert at smooth talking people into his confidence, then fleecing them by way of marked cards, dealing off the bottom of the deck, shell games and the fine art of duplicity.

Now on his own, the dapper-looking man with a black mustache, gray eyes, a flashy but false smile and a disarming southern drawl had learned to carefully select his marks. He steered clear of established residents, but concentrated on going after the contents of visitors' wallets. The thicker the wallet, the better.

Smith became the kingpin of crime in Denver, paying off the law and organizing a syndicate to extort all they could by illegal means from the people who roamed the streets of the Mile High City. When he heard of a silver strike in Creede, he took a couple of his henchmen there. The rush that brought scores of would-be miners to Creede wore out after a time and Smith sought greener pastures in the new mining town of Leadville.

It was in Leadville that Smith earned the "Soapy" moniker. He took a suitcase to a busy corner, set it up on a stand, then pulled out several bars of soap. He made a show of wrapping various denominations of currency around bars of soap, which cost him a nickel, then folding a wrapper around the soap, hiding the bill. Pulling more soap from the case, he shuffled the lot. By this time a crowd was beginning to appear on the sidewalk. The curious on-lookers asked how much he charged for the soap. "Only a dollar," he said as he continued to shuffle the bars around on top of the suitcase.

"I'll take one," a well-dressed man said as he pushed through, waving his dollar bill and taking a bar. Quickly removing the wrapper, he gleefully shouted, "Hey, I got a twenty-dollar bill!" At that, more in the crowd reached for their wallets. After a few other customers were disappointed with finding only a bar of soap inside the wrapper, another purchaser loudly proclaimed his good fortune at finding a five-dollar bill inside.

Of course, both "lucky winners" were in cahoots with the character who played on the greed of the gullible, turning the soap and their currency over to Smith when they were out of sight of the hopeful but disappointed customers. Smith became so adept at the scam that a growing list of admirers gave him the nickname that stuck for the rest of his life. At that point he had no idea just how short that period would be.

The first hints that some gold was being found in the far north brought Smith to Alaska in 1896. He went to Dyea, a short distance north of the camp that was to become Skagway. One suitcase held his clothing, another was filled with bars of soap and wrappers. In his wallet were a few bills of varying denomination. He set up his little stand and opened the suitcase, but he found the prospectors there were a bit more suspicious than his Colorado customers. When someone noticed that the "winners" began to look familiar despite attempts at disguises, a miner's meeting was called. By near-unanimous vote, Soapy was invited to leave on the next boat. Satisfied with what he had learned about the soon-to-be boomtown and happy that he had bilked the suckers out of enough money to pay his way south, he made plans to return.

In Seattle, Smith ran into a former police officer who had been on the take in Denver. They reminisced about old times, sharing such stories as the one about how Smith talked his way out of a charge of fraud. Soapy, it seems, was hauled in for defrauding two visitors of fifteen hundred dollars. In court and under oath, Smith testified that his Tivoli emporium was not a gambling house but an educational institution. Its purpose, he swore, was to "cure" people of the gambling disease by showing them that the odds were always with the house. "Why," he insisted, "we even warn them that they need to be careful. There's a sign right over the front door." The defendant failed to explain that the sign's message read, "Caveat Emptor," a Latin phrase that only a priest or a lawyer was likely to translate as "Let the buyer beware."

He was also known to boast that he was not a gambler, saying that a gambler is one who risks his money on the chance of being a winner. Soapy admitted that he left nothing to chance, that when he put up his money it was only on a sure thing. And he knew how to make it sure.

Soapy invited the lawman to go back to Alaska with him the next spring, boasting that he was going to run the new town of Skagway and promising to make his cohort the police chief. The man declined.

Agreeing to accompany Smith, though, were five former associates. Each had special talents that their boss would soon employ.

"Reverend" Charles Bowers was a saintly-looking man wearing a cleric's collar who was well-versed in the Bible and could offer solace to the sad and listen in seemingly sincere sympathy to their innermost secrets. He was adept at getting information from misfortune's survivors who while being comforted freely disclosed their financial condition. If they had gold or other items of worth being taken back to relatives of a lost gold-seeker, Bowers would see that someone would relieve the mourner of the valuables before they boarded a southbound ship. While he could speak of the glories of the Hereafter, he was completely steel-hearted and had no compunctions about being the instrument by which one would make an early trip to meet his Maker.

George Wilder was the advance man, going ahead of the others to prepare the people of Skagway for the arrival of important businessmen who would do wonders for the commerce of the new town. He dressed well and was able to pass as a wealthy stockbroker, being perfect in the role of letting someone in on a "sure thing."

Cyd Dixon looked the part of a playboy and world traveler. Having come from a wealthy family, he had made the mistake of being led into the clutches of the poppy's elixir. His opium habit made it necessary to get money, lots of money, any way he could. He fit perfectly into Smith's master plan.

"Slim Jim" Foster was a strong, baby-faced young man with blonde hair, a winsome smile, and the ability to work his way into anyone's confidence. Once established in Skagway, he met incoming steamers, offering to help passengers with their baggage. He asked their intentions, offering to steer them to whichever of Smith's enterprises would be most likely to extract the most of their money.

"Red" Gibbs could be counted on to do whatever was needed to further the cause. Smith's desire, no matter what was involved, was his bidding.

With these trusted and proven men as the nucleus of his organization, Jeff Smith was on his way to create an empire in this new town. He had correctly surmised that Skagway would be the gateway to the Klondike. The groundwork was laid. His plan was set. He calculated every step required to put it in motion and lost no time in getting it moving once he stepped off the boat.

The man himself was not above setting up a table alongside one of the trails leading over the mountains to the streams that were so rich in gold. Three walnut half-shells were the tools of his trade. When a sucker paid his dollar and took the chance that he could keep track of the one under which a pea was placed, the quick hands and slender fingers of the former soap-seller shuffled the shells. The pea was never under the shell picked by the bettor. If challenged, Smith would lift one of the other two shells to disclose a pea. The bilked stampeder didn't know that Smith only bet on sure things. He also was unaware that Soapy was a master at sleight-of-hand, quite capable of palming a pea.

Smith's talent was even greater in dreaming up ways to extract money from the uninformed.

What, Soapy surmised, would be one of the first things a newly arrived prospector would want to do? Obviously, it would be to let the folks back home know he was safely ashore and preparing to attack one of the passes to the land where dreams come true. The solution, of course, was to provide a telegraph service. A building was rented and a sign erected to identify the location as a telegraph office. The clerk wrote down the newcomer's message, took the five-dollar fee charged for a telegram going anyplace in the world, and handed it off to the "telegrapher" behind the screen. Relieved, both figuratively and literally, the customer was asked where he would be staying in case there was an answer. There usually was an answer within a few short hours, delivered to the customer in return for another five dollars.

The telegraph office was a big hit with newcomers who flocked to the counter with their five-dollar bills in hand. They were happy to learn that their message had been received by relieved friends or relatives and that they sent well-wishes for the journey over the passes—willingly handing over another five-spot.

There was one catch. The customers did not realize that there were no wires running to the office. There would be no telegraph service connecting Skagway with the outside world for another three years. The money received for messages thrown in the trash and answers that were pure fiction was clear profit.

Another business in high demand was a tourist information office where maps, supply lists, information of use to an inexperienced prospector, and other material helpful to a cheechako could be acquired. Maps and pamphlets were quickly made up and printed in the shop of one of the two newspapers that had sprung up in the camp. The printing was done at a nice discount because Smith quickly ingratiated himself with the plant's owner. The money that exchanged hands between them was well spent because the smooth-talking entrepreneur desperately needed good publicity—and was willing to pay to avoid any that was bad.

Saloons were the mainstay of the town's business district and Smith opened one, calling it Jeff's Parlor. A back room was set up for gambling. The odds there were definitely always with the house. No customer was able to boast about beating those odds. If he were ready to leave the establishment with a bulging poke, there was a good chance that another customer would snatch the poke and run out the back door. Other customers would immediately give chase, in doing so creating a jam that blocked the door long enough for the culprit to disappear.

One enterprise that was not successful, however, was the bank that Smith established. Depositors wanted receipts and assurance that their gold was safe. Frequent unsolved robberies scared them away and there was no reason to keep the bank open. It was closed, giving way for someone to open an honest institution.

Smith's enterprises grew. He hired, as shills and enforcers, prospectors who did not have enough money to cross into Canada and could not find legitimate employment. He paid the most productive of his business associates half of their ill-gotten receipts. There was plenty left for Smith to become wealthy.

His young associate, Slim Jim Foster, was highly paid. He was the kindly person who helped new arrivals with their baggage, then steered them to one of Smith's enterprises where they would be relieved of their bankrolls.

There was no law in Skagway. When a deputy United States Marshal was assigned there, the man quickly found Soapy to be a more generous employer than was the government. If a customer at Jeff's Parlor complained about being the victim of a strong-arm robbery, the hapless individual was told that nothing could be done because he was unable to positively identify the culprit.

On the rare occasion when an unhappy newcomer raised a fuss, the magnanimous and pubic-spirited Jeff Smith was sympathetic enough to pay the person's fare back to civilization Outside. Quieting the victim with a ticket home carried less risk than killing him and raised less public indignation.

Because the thievery was directed at strangers bound for or returning from the Klondike, most of the permanent citizens of Skagway were not even aware of the skullduggery engaged in by Smith and his gang. To help Soapy's image, the friendly newspaper editor frequently published accounts of Soapy's generosity. When a church was built, Smith was praised for contributing much of the money needed for its construction. He contributed to the hospital and the fire department fundraisers and to other civic endeavors. He was patriotic, championing the country whose free enterprise system allowed him to become wealthy. On the Fourth of July, he decorated the main street and, astride his gray horse, led the parade. At the 1898 celebration he sat on the platform with Alaska Governor John Green Brady, who gave a speech and in the process lauded the town's leading citizen, the grand marshal of the parade, Jeff Smith.

Four days later, Jefferson Randolph Smith would lie mortally wounded on the wooden approach to a warehouse.

For some weeks the increasing incidents of crime had been causing concern among the legitimate business people of Skagway. On February twenty-fifth, a Seattle newspaper ran a story stating, "Lawlessness in Skagway is beyond description. Jeff Smith and his gang are in full control. Holdups, robberies and shootings are part of the routine. Eight dead bodies were picked up on White Pass on February 15." It continued to detail incidents that had come to light.

One incident of thievery started the final chain of events that culminated in Smith's death. A prospector named J. D. Stewart was leaving the Klondike, taking with him a poke containing twenty-eight hundred dollars in gold. He was headed for his home in British Columbia. Although warned of Smith's operation and told to deposit his gold in the safe at his hotel, he unfortunately fell victim to the smooth talk of Old Man Tripp, the Moses look-alike, and "Reverend" Bowers. They proposed that he take his poke to Smith's assay office and refining facility where he could get more cash for his gold than he would at an Outside brokerage. On the way there, Stewart was detoured to Jeff's Parlor and into the back room. There, a stranger grabbed the poke and took off through the back door.

Stewart complained loudly to the deputy marshal but after admitting that he could not positively identify the thief, was told there was nothing that could be done. Not satisfied, Stewart carried his complaint to the newspapers and told his story to everyone he met. The incident caused a great deal of talk. One businessman argued that if this could happen to one of the first people leaving the Klondike, others would choose to take the longer but safer steamboat ride downriver to St. Michael.

A major obstacle to bringing law and order to Skagway was the fact that Smith had organized a small army of enforcers who were well armed. He also had formed a network of informants, many of whom were on the borderline of legitimacy. His henchmen attended church, appeared to be honest workers and shopped at the stores of the business people who had become concerned.

A vigilante committee was formed in an attempt to gain some control over the criminal element. One of the primary leaders was a man named Frank Reid, the engineer who laid out the townsite and designed Skagway's sewer system and port developments. Reid was in his mid-fifties and had worked his way north after graduating from the University of Michigan. He had fought Indians in what was to become Nevada and Oregon and headed to Skagway as soon as he heard of the Klondike strike.

Joining Reid as leaders of the Committee of 101 were Major J. M. "Si" Tanner and Captain J. L. Sperry, both of whom also had been Indian fighters and were former police officers. At a meeting of the committee, it was decided to ask for federal troops to be brought from Dyea and that martial law be declared. The committee was reluctant to go against the Smith organization, their hesitancy strengthened when Smith strode into one meeting, a revolver in each hand, threatening to shoot if the meeting did not break up. It did and Smith left, a smile on his face.

Until this point more interested in protecting his image as an upright citizen than in confrontation, Smith became obstinate. His ego possibly more inflated by his recent appearance with the governor, he made it clear that he was not going to give up control of Skagway.

A meeting of the committee was called for July 7 but broke up when Smith's people took over and nothing could be accomplished. Another meeting was called for the next night in a warehouse on the Juneau Dock in the Skagway waterfront. To keep Smith's people away, four men were assigned to guard the dock. Reid was at the forefront.

Smith approached the dock and was told by Reid that he could not pass. Smith objected with a curse. Armed with a Winchester 30/30 slung over his shoulder, a derringer up his sleeve and a revolver in his pocket, the confidence man swung the rifle around, only a few feet from his adversary. Reid drew his Colt forty-five and grabbed Smith's rifle. Both weapons went off simultaneously. Smith dropped dead, shot through the heart. Reid was hit in the groin and lay in agony for twelve days before passing away on July twentieth.

Smith's gang quickly broke up after the shootout. His top lieutenants were rounded up. Slim Jim Foster was taken by vigilantes and a noose put around his neck. The lynching attempt ended when Tanner, the Committee of 101 co-leader, stopped it with a threat that sank in. Old Man Tripp and "Reverend" Bowers had found refuge in the upper floor of a waterfront building but were drug out. Frontier justice was averted when soldiers from Dyea arrived and took Smith's lieutenants in custody. They were later transported to Sitka to go before a grand jury. Others of the gang who could be identified were sent to Seattle and told never to return.

A search of Smith's belongings turned up Stewart's stolen poke, five hundred dollars in gold missing.

When the scourge of Skagway, Jefferson Randolph Smith, was to be buried, ministers of the Baptist and Methodist churches refused to officiate. The town turned to the Presbyterian, Sinclair, who had murmured thanks upon seeing the mortally wounded man lying on the dock, his life blood turning the boards red. At the church there was only one mourner present: Smith's girlfriend who stood alone in the doorway of the church Smith had helped fund. The only other occupants were the minister and the vigilante he had invited to be there in case anyone wanted to vent his outrage at the service. Sinclair's funeral sermon was based on the thirteenth chapter of Proverbs. Several verses would have been appropriate. "The light of the righteous rejoiceth: but the lamp of the wicked shall be put out" is one. The verse emphasized by Sinclair, though, was "Good understanding giveth favor: but the way of transgressors is hard."

Smith, the man who temporarily controlled the town as he had predicted a year earlier, was buried outside the cemetery, away from its assortment of better souls who came to the end of their trails in this rugged frontier town. A small headstone marked his grave, eventually to be replaced after souvenir hunters made off with all but a small piece now held as a reminder of the town's heyday as the gateway to the Klondike. One mystery remains: the name of the anonymous donor who sent fifty dollars each year to Harriet Pullen, owner of the Pullen House hotel, as compensation for keeping Soapy's grave clear of weeds.

At Reid's service, on the other hand, the entire town turned out to honor the man they saw as having "saved the honor of Skagway." The hero was buried in the town cemetery, his final resting place marked by a towering headstone proclaiming the esteem in which he was held.

Sadly, it is the scoundrel Smith whose final resting place today is most sought out by visitors. His watering hole, Jeff's Parlor, is maintained as Skagway's top tourist stop.

Jefferson Randolph "Soapy" Smith, left, controlled Skagway's underground while putting on a face of civic good. His Jeff Smiths Parlor, below, was headquarters for a band of confidence men who used many schemes to extract gold from unsuspecting stampeders. He died in an exchange of gunfire with vigilante Frank Reid on July 8, 1898.

Photos courtesy Library of Congress

George W. Carmack credited with gold discovery

Before we leave the Canadian portion of this account of the Great Gold Rush, there should be given attention to the man who is credited with starting it all. Just as was the man himself, his claim to be the discoverer remains controversial. Canadian officials now recognize several co-discoverers.

George Washington Carmack was born on September 24, 1860, in Costa-Contra County, California, to Christopher MacKenzie Carmack and Anna Siemens. His mother died when the boy was eight and his father three years later, leaving him an orphan at eleven years of age.

Carmack joined the United States Marine Corps and was serving aboard the *USS Wachusetts* in 1882 when he got word that his sister, Rose, was sick. Refused his request for leave, Carmack reportedly said "goodbye" to the Corps and jumped ship. He had served briefly in Alaska and returned there as an apparent fugitive from desertion charges in 1885. Engaged in trading, fishing and trapping, he traveled extensively along the rivers of Southeast Alaska and often crossed over into Canada.

Two years after returning to Alaska and roaming the Yukon, Carmack took as his common-law wife the daughter of a chief in the Tlinkit Crow clan. A widow, her husband and infant daughter had both died of influenza. Her Tlinkit name was Shaaw Tlaa, but after joining Carmack she took his surname and the given name Kate. The Carmacks were often accompanied by her brother, Skookum Jim, and her nephew, Dawson Charlie.

In August of 1896, George, Kate, Jim, and Charlie were fishing on the Yukon when they encountered Robert Henderson, a prospector who had found some color in creeks in the area. Following the creed of the north, he suggested that Carmack stake a claim on nearby Gold Bottom Creek, which fed into the Klondike River.

Although he had discovered a coal deposit in an area that later was named CarCross in his honor, the old-timers who mined at Fortymile and the prospectors who radiated from that place said Carmack was not very interested in mining for gold. He seemed to prefer an easy-going Native lifestyle, hunting and fishing.

As a matter of fact, the white men in the area found it difficult to know what the man stood for. He generally was known as "McCormick." Most of them had little use for him due to his habit of exaggerating everything. No matter the subject, he usually had a story to top that just told by someone else.

What is known is that Carmack and his group did not try Gold Bottom but went a little farther upriver and stopped to fish on Rabbit Creek, another stream that fed into the Klondike.

Several stories describing the find have been recorded. One version is that while the men were sleeping one afternoon, Kate was getting water from the creek when she saw flecks of gold in the gravel. Another is that it was Skookum Jim who made the discovery. All the versions agree that their first pan showed a quarter of an ounce of gold, worth four dollars. To the uninitiated, the excitement generated by that amount of color showing in the sands at the bottom of the pan is explained by the fact that eight *cents* worth of gold was considered a good prospect, ten cents a definite find. This was a bonanza, a name later given to the creek where the gold was found.

Whatever the truth, Carmack staked a discovery claim in his own name and headed by a roundabout route to Fortymile to record it.

He did not tell Henderson of his discovery, a fact that brought about a feud between the two men. Again, controversy and differing versions cloud the facts. One version is that Carmack and his relatives were angry because Henderson had said he did not want any Natives filing on "his" creek, using derogatory terms in saying so, and ignored him while telling others of their find. Carmack's supporters say the discoverer was just too excited and forgot to go back to find his benefactor.

Whatever the real story of the discovery, it is not disputed that Carmack filed a claim and the discovery made that day is what started the last great stampede. The four dollars found in the pan of gravel from Rabbit Creek was but a tiny hint of the billions that were to come from the creeks of the far north.

Carmack and Kate departed Alaska in 1900. They moved in with Carmack's sister Rose, who had survived the illness that caused her now-wealthy brother to desert, near Hollister, California. Continuing to seek adventure, Carmack abandoned Kate and their daughter. Unable to show that she was his legal wife and entitled to alimony, Kate returned to Carcross. She worked as a seamstress, having previously followed that line making warm clothing for the miners. She died of influenza in the global outbreak of 1918-1920.

In 1900 George Carmack married Marguerite P. Laimee in Olympia, Washington. They moved into a large home in Seattle. She convinced him to invest in real estate and they prospered. George continued to prospect in search of another bonanza and died in 1922 while working a new claim. He was 62.

One last controversy involving the man erupted when Graphie Grace Carmack, the daughter of George and Kate, and George's sister Rose went to court to challenge Marguerite's appointment as administratrix of the estate of the dearly departed discoverer. The suit was settled out of court, the details unreported.

George Washington Carmack and his common-law wife, Kate (Shaaw Tláa), are now considered as co-discoverers of the Klondike gold. Photographer unknown

Robert Henderson upset over Carmack recognition

Robert Henderson died in bitter obscurity thirty-six years after suggesting that another man check out a mining prospect on a nearby creek. The lack of recognition of his role in the discovery of gold in the Klondike region of Canada's western edge was a disappointment carried to the grave. It was an issue that stretched across oceans. It caused tensions across national boundaries. It raised racial overtones. It caused monuments, with plaques whose wording was agonized over for months, to be destroyed and replaced multiple times.

Henderson was born in 1857, the son of a Scottish lighthouse keeper at Big Island, Pictou County, Nova Scotia. He developed a fascination for gold at a very young age, left home and signed on as an apprentice with a sailing ship at age fourteen. He traveled distant lands, prospecting in Australia, New Zealand and California in a never-ending pursuit of the yellow metal. He once planned to travel to Alaska, but missed the boat in Portland, Oregon. He returned to Nova Scotia, fell in love and married. In 1894 he went with two partners to the Yukon. The partners gave up, but Henderson stuck it out.

Stopping at Joseph Ladue's trading post in the spring of 1896, Henderson was urged by Ladue to prospect farther upstream where good color had been found. The trader grubstaked the prospector who paddled a moose-hide canoe upstream to the Indian River. On one creek, he panned eight cents worth of gold—an amount considered in those days to be a good prospect. He named the creek Gold Bottom and staked a claim. Instead of setting about to work the ground, he continued prospecting, but as was the custom, told others he met about his find.

On a fork of Gold Bottom he met two other prospectors and told them about his luck, only to learn that they had staked a claim on a prospect that showed thirty-five cents to the pan. Henderson staked a claim alongside theirs and continued on to the Stewart River where he staked a third claim. It was there that he met a group moving down the Klondike. When he told them of the two prospects, they informed him that "McCormick" had discovered a prospect yielding a quarter of an ounce of gold—four dollars—on Bonanza Creek and they had staked there. Henderson did not recognize the name, but when they pointed to a rounded hill and said it was just on the other side, he knew it was the place where he had directed Carmack. He had expected Carmack to let him know the results of his prospecting and was angry not to have heard.

It did not occur to Henderson that Carmack and his Indian companions had been offended during their earlier encounter. Carmack was accompanied by Skookum Jim, the brother of his common-law wife, and her nephew, Tagish Charlie. Carmack's reputation as a man who preferred to live a Native lifestyle was known around the river. Henderson had made the statement that he did not want any Natives filing on "his" creek. As it turned out, it probably was Jim who actually found the gold while Carmack and Charlie were fishing. Carmack, though, put his name on the claim of discovery, September 24, 1896.

Henderson's next disappointment came when he went to Fortymile to file his claims. He was told that the law had been changed. Under the new requirement, a prospector could file on only one claim. Since "Old Man" Hunker and his partner had earlier filed on Hunker Creek, Henderson could only have one standard size claim rather than the double-size of a discovery claim.

Henderson decided to take a steamer downriver and head Outside. His bad luck continued when the boat was frozen in at Circle City and could go no farther. Held over, a leg injury became infected and Henderson needed medical treatment. With no money and no gold produced, he ended up selling his claim for three thousand dollars.

When the river opened and he could continue on to St. Michael, Henderson had but eleven hundred dollars left of the three thousand.

The dark cloud under which Henderson had been cursed to remain continued. His gold was stolen from his stateroom. He had nothing to show for years of prospecting in what was destined to become one of the richest mining areas ever. On top of that, Carmack was being celebrated as the discoverer who started the great gold rush and no credit was given to the man who recommended the place for him to look.

On the way home to meet his wife, Henderson ran into Tappan Adney, a writer for *Harper's Weekly* who was covering the stampede for that publication. The hapless prospector removed a badge from his jacket and handed it over to the writer. "You take it," Henderson said. "I'd probably just lose it, too." Adney wrote that he looked at the familiar insignia of the Yukon Order of Pioneers and recognized the irony when he read the motto imprinted around its edge: “Do unto others as you would be done by.”

Henderson felt he had been cheated not only of recognition but of money when the government changed the filing laws. His claim was staked before the change was made, and at that time there had not been the sixty-day filing requirement.

The argument took wings. Because Carmack was an American and Henderson a Canadian, national biases entered the picture. Editors on both sides of the border joined in. Then a third element was raised: members of Canada’s First Nation said that Skookum Jim, whose Tlingit name was Keisch, was the real discoverer but had been pushed aside by racial discrimination.

Canadian Robert Henderson gained recognition as co-discoverer after first being denied. He died embittered because of the slight.

Photographer unknown

Surveyor William Ogilvie played major Klondike role

A man who had just turned forty-one left his family and a comfortable life as a surveyor in Ottawa, Ontario, to take an honored place in the history of the Klondike. Sent there in 1887 by the Canadian government to survey the area, he was to become unofficial judge, peace officer, friend to miners and traders, and later to replace Gold Commissioner James Morrow Walsh. In the latter situation he was faced with cleaning up what one historian described as "a cesspool of politics and corruption." His major professional achievement was locating the boundary between Canada and Alaska, settling an argument that had gone on since 1825. His legacy is having several landmarks named in his honor and in writing a history of the Gold Rush. His book, "Early Days in the Yukon," published posthumously in 1913, is a definitive history of the Gold Rush.

William Ogilvie was born in Ottawa on April 7, 1846, of a father who was an immigrant from Ireland and a mother who was born in Scotland. He attended Ottawa College where he studied science and mathematics. He had an intense interest in geology and botany, subjects that were to show clearly in details recorded in his reports and publications. He apprenticed to Ottawa surveyor Robert Sparks. He married Sparks' sister, Mary Ann Sparks, on March 8, 1872, and the couple had three sons. Little is recorded of his family other than that his youngest son died while Ogilvie was in the Klondike and his eldest son went north with him on a later trip there. His final trip was as president of a dredging company of which the elder son, Morley, was chief engineer.

The surveyor's work was widely recognized. He was meticulous and relied on exact science—at it was then known—to establish base markers from star sightings. It was this reputation that caused the government in 1887 to select him to travel to the Yukon area to map routes and rivers that were already getting attention from traders, trappers and prospectors who moved about the continent's longest river. The Yukon's headwaters were in a wilderness area in the northwestern part of the dominion and wound its way two thousand miles to the Pacific. One-third of the river's route was in Canada while the larger portion was in Alaska, a United States possession acquired from Russia in 1867. That distinction was to become a major factor in Ogilvie's work on his second trip north.

Ogilvie was assigned to a six-person party under Dr. George Mercer Dawson. The Dawson party surveyed the Chilkoot Pass, the Porcupine and the Yukon rivers and did some work in connection with identifying the location of the one hundred forty-first meridian west. It was that line that was the border between Canada and Alaska—a line that fluctuated based on indefinite measurements.

Arriving at Dyea, the survey party contracted with a team of one hundred twenty Chilkat residents to pack their seven tons of material up the steep pass. Among the group were three people who were destined to make their own mark on history: George Carmack, "Skookum" Jim Mason and Kaa Goox, a Tagish Native, the nephew of Carmack's wife and later to be known as "Dawson Charlie" Henderson. Ogilvie was very favorably impressed with Mason and spoke highly of him in his book. The word "skookum" is translated as "strong," fittingly applied as Ogilvie wrote that Jim packed one hundred and fifty-six pounds of bacon in one load to the summit of the pass. The pass rises three thousand feet above sea level, the last three-quarters of a mile so steep that one must use his hands to claw his way forward. Ogilvie was to be known as completely fair in his dealings, but as was the case for many of the white interlopers into the north, his impression of the Natives was highly prejudiced. Jim was employed by the surveyor on several occasions and was praised for his industriousness; but that praise was sullied by the intended compliment that Jim was highly intelligent "for an uneducated savage." Ogilvie also distrusted the Natives because of the proclivity of some to appropriate items for their own use and for their habit of asking for more in the way of payment while in the midst of the job, especially during the climb up the pass. Their taste for the liquor that the newcomers provided was also a point of contention.

After completing his surveys of the Chilkoot and Porcupine River, Ogilvie went downstream, marking the path of the Yukon until reaching the new mining camp at Fortymile. From there he made his readings from "star shots" to establish a base mark, then proceeded to find the meridian that was the national boundary established in a treaty between Russia and Great Britain, of which Canada was a dominion, in 1825. The purpose of that treaty was to determine the locations where each country's trading firms could operate—and from whom they could collect tribute. Just where the meridian crossed the Yukon was constantly under question due to competition between the Russian-American Company and the Hudson's Bay Company. It was previously surveyed by both Canadians and Americans in 1869, 1877 and 1883, the latter one by United States Army Lieutenant Frederick Gustavus Schwatka. Ogilvie was to find that all the earlier surveys were incorrect and Schwatka's mark was twelve miles off the actual line. Hudson's Bay trading posts west of Ogilvie's line were expelled and American traders moved in. Some mining claims were found to be in a different nation than had first been thought.

The settlement of Eagle was found to be located in Alaska, rather than on the Canadian side of the line as originally thought. An Army post was quickly built there.

Three principals among the American traders were Jack McQuesten, Arthur Harper and Al Mayo, all of whom were influential along the entire river and destined to become close friends of Ogilvie. While the Hudson's Bay traders had pulled back from Alaska, the American firms of Alaska Commercial Company and North American Trading and Transportation Company operated over the length of the Yukon. Their stories appear later in the section that deals with the Alaska portion of the rush.

Ogilvie's principal work in determining the boundary completed, he headed home after two years in the area. He was reluctant to leave the many people with whom he had become friends but was also anxious to get back to his family and to turn in his voluminous reports.

Those reports gave his notes and conclusions on the boundary line. Of equal importance were his reports on conditions in what was later to become the Yukon Territory. During his time at Fortymile he had met many of the men who had been in the area for some time. The majority of them were Americans who had complaints about the differences between Canadian and American mining laws. They preferred the American way of staking and filing claims and their size. Most Canadians agreed on major points, particularly saying that claims should be larger than allowed under laws of the dominion. In his book, Ogilvie said the views of the miners varied considerably, but "The best I could do was to hear as patiently as possible the views of all," and gave accurate summaries in his report. Many of his recommendations were followed, especially in changing the size of claims. The Canadian regulations for placer claims as applied to the Yukon and its tributaries were thenceforth to be five hundred feet in length along the creek and from rim to rim of bedrock.

In 1893 the surveyor was sent to the west again, this time to survey the boundary around the Alaska Panhandle. Part of that assignment was to investigate a report of Native unrest received from Superintendent Charles Constantine of the North West Mounted Police, a report that Ogilvie found to be without basis. Returning to Ottawa in March of 1895, Ogilvie learned that his youngest son had died in January. He blamed Constantine for causing him to be away from home in that time of need. A rift between the two continued, with Constantine angered by Ogilvie's successful recommendation that a more limited police force was needed than the superintendent wanted.

Going back to the Yukon in 1896 to make further examinations in preparation for a commission to resolve the boundary question, the surveyor completed that work and dispatched his report by mail. Awaiting instructions from Ottawa, he was asked by Joe Ladue to survey the township he was establishing near the confluence of the Klondike and Yukon rivers. Despite Ladue's desire to name the town for himself, Ogilvie named it Dawson, in honor of the former expedition leader who had distinguished himself.

Wintered in by an early freeze, Ogilvie was asked by mine owners to help settle local boundary disputes. The claims had been measured haphazardly and their lines were in dispute. Some newcomers filed competing claims on what they said were excessive portions. A miner's committee had attempted to reach a settlement, but as unqualified surveyors their decisions were discredited. Ogilvie said he would agree only if a majority of claim owners requested his service. Petitions were quickly gathered and signed by a substantial majority.

It was an enormous task. The procedures in mining regulations called for a stake of least four feet high be driven in at the downstream corner of the claim and marked with the claimant's name and pertinent data. Corner Two was to be five hundred feet *in a direct line* upstream, Corner Three across the stream and Corner Four five hundred feet downstream. In their haste to record the claims, most prospectors were not careful in taking measurements, just stepping off the distance. Some used ropes, trusting the length to be as marked when purchased. Some of the claimants misinterpreted the regulation, pacing the distance along the edge of the creek; where there were bends in the stream, those claims were short. A rope that was shorter than the perceived length resulted in a claim that was smaller than allowed. Failure to place the third stake at ninety degrees resulted in a pie-shaped wedge.

As a government official, Ogilvie with only one exception drove in his stakes where they should be; the discrepancies were left for the owners and would-be filers to discover.

The lone known case where Ogilvie pointed out a discrepancy was on *Five Eldorado*, owned by Clarence J. Berry, destined to become one of the most successful of the Klondike miners. Berry's workers had thawed ground and dug shafts to bedrock, placing the dump of gold-bearing material on the upper end of the fifth claim in the string upstream of *Discovery*, to be run through sluices in the spring. It was late in the day when Ogilvie was ready to drive his stake at the uppermost corner of the claim. Berry was standing by, anxiously awaiting the placement of the stake for Corner Two. Rather than drive the stake, the surveyor dispatched his crew to another claim, then turned to the owner and asked if it were not time for supper. Berry looked at him and asked if something were wrong. Ogilvie again asked if were time for supper and started towards Berry's cabin. On the way there, the owner again asked what was the matter. The surveyor responded that they should go on in for supper. Once inside and Berry's wife Ethel asked to put on a plate for their guest, Ogilvie quietly explained that *Five* was forty-one feet, six inches longer than it should be. Realizing that his entire winter's product was stacked on what would soon be known to be open and subject to claim by someone else, Berry anxiously asked what he could do. Ogilvie asked if Berry knew someone he could trust who did not already have a claim. In his book, Ogilvie said Berry came up with the same name that had occurred to the surveyor. Without another word, the claim owner put on his warm garb and rushed three miles upstream where his friend was working. Ogilvie waited at the Berry cabin. When the two men returned, George Byrne joined the two other men at the table and quietly asked about the procedures for staking a claim. Ogilvie wrote that even Ethel Berry was not aware until much later of the predicament her husband was in. Byrne drove stakes forty-one feet, six inches up from Berry's and the next morning filed on that fraction. He then transferred the claim to his friend. When sluiced, the cleanup from that dump was estimated to be worth one hundred thirty-six thousand dollars.

While Ogilvie was prohibited from filing on claims, that did not apply to members of the Mounted Police or other government employees.

Dick Lowe, a chainman on Ogilvie's crew, saw that a wedge-shaped fraction measuring eighty-six feet at its widest point, had been disclosed in the survey. He promptly staked the piece, thinking it could be sold profitably. No one was interested, however, in buying such a small claim. Lowe decided to mine the fraction himself. His first shaft found nothing. The second, however, produced forty-six thousand dollars in just eight hours' labor and by the time it was worked out was worth more than a million dollars. That fraction was considered the most valuable per square foot of any Yukon claim.

Anxious to get back to his family in Ottawa, Ogilvie finally was able to wrap up his work. He took a steamer to St. Michael and there boarded the *SS Excelsior.* That same ship landed the Berrys and many other Klondike miners in San Francisco on July 14, 1897. The gold they carried caused a stir in that port, but not nearly to the extent raised by the *Portland's* arrival in Seattle three days later to spark sensational headlines.

Back at his desk in Ottawa in September, Ogilvie wrote a guide book on the Yukon, intended to correct some of the misinformation being promulgated by get-rich-quick authors whose works were more imagination than fact.

Three months later, now fifty-one and after enduring numerous trips to the Yukon, Ogilvie was once more called on to serve his country. Walsh, the former mounted police major appointed as commissioner of the Yukon, had failed in his job. There were charges of corruption and ineptness, leading to his being relieved. Ogilvie headed a commission investigating the various allegations leading to Walsh's dismissal but politics and greed kept interfering with justice. He finally resigned in the summer of 1901, citing ill health.

Going downriver on a small steamer, Ogilvie met Miss O. P. Richardson who was headed for Fort Seward in Haines, Alaska, to meet her brother, United States Army Captain Miles P. Richardson. The captain was a road-builder and surveyor who went to Alaska in 1899. At St. Michael, Ogilvie and the Richardson woman were in a small boat that capsized in Norton Sound. Ogilvie was credited with keeping her from drowning by holding her up until rescuers plucked them from the icy water.

Mystery surrounds the family life of the famed Yukon surveyor. What happened to Mary Ann is not known. Nothing is known of Miss Richardson, who became Mrs. Ogilvie at her hometown in Paris, Texas, on April 15, 1903. It was she who wrote the preface to Early Days in the Yukon, signing it simply as O.P.R. Ogilvie. The book, published a year after the author's death, was dedicated "To My Wife." It is filled with dry humor and contains many humorous anecdotes.

Ogilvie died in Winnipeg on November 13, 1912, at the age of sixty-six.

His legacy is in many geographical locations bearing his name. He is listed in the Yukon Mining Hall of Fame. He was judicial, fair, dedicated to his work, meticulous, ever faithful in getting it right, and the possessor of a good sense of humor.

The Yukon Territory and Alaska are so bound together by a common interest that Canadians and Americans have joined their hands in a combined effort toward the best development of the country.

While the International Boundary is a broad line through the wilderness, men in this land of the midnight sun are prone to forget nationality in remembering the tie of Arctic Brotherhood.

--O. P. R. Ogilvie in Preface to *Early Days in the Yukon*

Part II

The Boom moves downriver

The Klondike Stampede that brought so many people up the river and over the passes wound down soon after the dawn of the Twentieth Century. Some abandoned their claims as "worthless." Others sold out to dredging companies such as the one that brought bankruptcy to Gene Allen of the Nugget Express due to delivery of heavy equipment but at a cost far exceeding his contract price.

An even bigger cause of the exodus of individuals who mined in the Klondike was the discovery of gold at Cape Nome. Word that the "Three Lucky Swedes" hit it big on Anvil Creek and, soon after, that nuggets could be picked up right on the beach swept through the area at the headwaters of the Yukon. Those who did not head back to their homes Outside moved to the west, to the new town of Nome.

As at frontier Skagway, law and order were not present at the western edge of Alaska's vastness. That meant the area was fair game for all kinds of practitioners of chicanery. Nome was to have its own cast of characters, some even more colorful than those at the other end of the river.

Actually, Nome is not located on the Yukon River. It lies several miles north and west of St. Michael, the seaport at the delta that was the mouth of the Yukon. The Seward Peninsula is a large appendage jutting into the Bering Sea on the west coast of Alaska. Norton Sound is on its south side, Kotzebue Sound to the north. Nome was incorporated in 1901 and became the business center for northwest Alaska.

Search for Alaska gold began before 1867 purchase

The first recorded discovery of the yellow metal was in 1849 when Peter Doroshin, a Russian explorer, found some gold on the Kenai Peninsula in southcentral Alaska. In 1861, "Buck" Choquette found gold on Telegraph Creek in Southeast Alaska, which sparked the Cassiar Rush in British Columbia. Ten years later, gold was found in Sumdum Bay in Southeast. In 1865, David Libby, a member of an exploration party surveying a route for a trans-oceanic cable linking the United States with Europe, found gold near where the discovery of 1898 was made. Later Libby established the Eldorado Mining District at Circle City, taking out seventy-five thousand dollars in gold. In 1880, Richard Harris and Joe Juneau found gold in creeks near the future Alaska capitol which would bear Juneau's name. Across the Gastineau Channel, John Treadwell found a vein of gold ore and began Alaska's first hard-rock operation on Douglas Island. Treadwell's operation was in full swing when the 1896 discovery was made in the Klondike region, over the mountains north and east of Juneau. Closer access to the latter fields was from Dyea, and then Skagway. Dyea, Skagway and Juneau are located on Gastineau Channel, a narrow arm along the Pacific Coast.

Alaska Native John Minook gains equal right to mine

John Minook, who developed several mines, was born alongside the Yukon River near its confluence with the Tanana in 1874, the son of Russian-Koyukon trader Pitka Pavaloff. Educated in mission schools—and by white prospectors who visited the area—he spoke English highly flavored with words not acceptable to the missionaries.

By the time he was twenty, he was considered to be a skilled miner and woodsman. He is credited with the initial discovery of gold on what now is known as Minook Creek, where the town of Rampart sprang up. Initially, the settlement was known by his name. Even though laws at the time prohibited Alaska Natives from owning claims, those who formed the Rampart Mining District in 1895 declared him to be an owner.

Minook and his children prospected over a wide area and mined extensively. Because of the situation existing at the time, little is known of their activities. Minook was inducted into the Alaska Mining Hall of Fame. In Minook's biography on the Hall's Web site, Thomas K. Bundtzen wrote, "Throughout his long career, John Minook was known as a very hard working and reliable prospector and placer gold miner. One issue of the Rampart *Whirlwind* described John as one of the most honest men in Interior Alaska; many accounts attest to his reliability."

William Haskell, in his book "Two Years in the Klondike and Alaskan Gold Fields 1896-1898" misspelled Minook's name as Munook and described him as "an old Indian" and "a Russian half-breed." He credited Minook and his son with "in a short time" during 1896 taking $3,000 in gold from a hole eight feet square and fifteen feet deep. Their find was not known until a steamboat stopped by. Word spread quickly and a stampede ensued, causing the Alaska Commercial Company to build a trading post nearby. Soon there were a thousand people in the camp which was named Rampart. Had there not been a Klondike discovery, Haskell mused, "it would have been enough to start a stampede" on its own.

In 1904 Minook applied for United States citizenship in order to clarify his right to own claims. Judge James Wickersham ruled that the question was moot inasmuch as Article Three of the treaty ceding Russian-America to the United States gave citizenship to people then residing within the possession.

Minook's lawsuit was far more complex and meaningful than the above statement indicates, however.

At the turn of the century, the status of Alaska's Natives was uncertain. Wording in the purchase treaty was not exactly clear. It excluded "uncivilized native tribes" from citizenship, but failed to define what that term meant. At the time, official census records listed several classes based on race: white, Creole, Native, or mixed-blood. In Outside courts, American Indians in some jurisdictions had to prove they were "almost white" to be considered for citizenship.

Being an American citizen was important to Minook.

A miner's court at Rampart declared him to be a citizen. But Minook's brother, Pitka Pavaloff, co-discoverer of the Circle mining district, was declared ineligible to own claims, losing his title to valuable ground. Minook's wife and children were mining in the Rampart area and it was important to establish their right to do so.

Wickersham pondered the question for a full year before settling on the treaty provision as the solution.

Official reports of gold production show that forty-eight thousand ounces of gold were recovered from one single claim on Minook Creek. Said to be worth a million dollars in those days, in today's market it would be about fifty-eight million. The Minook family continues to be prominent in Alaska.

Dog team preparing to leave Dawson bound for Valdez in 1901. Dawson City Museum, 1998.22.718

Vermont man details mining on two sides of border

William B. Haskell, a Vermont native whose wanderlust took him to Alaska in 1896, did not make a huge fortune nor did his success rank him among the most worthy of note in the annals of the great adventure. Upon his return to the States, however, he wrote a book chronicling in straightforward terms his experiences. Canadian historian Pierre Berton, considered to be one of the most reliable experts on the Klondike, wrote that Haskell's book was one of the best first-hand eyewitness accounts of the Gold Rush.

Haskell's book, "Two Years in the Klondike and Alaskan Gold Fields 1896-1898, A Thrilling Narrative of Life in the Gold Mines and Camps," was reprinted a century later by the University of Alaska Press. It provides a good description of the hundreds of people who were to chase the rainbow to the Yukon. The book gives a close look at the miners who worked at both ends of the fabled river that runs through the northern ribbon of gold.

The book was published in 1889. That year a dam gave way at Johnstown, Pennsylvania, killing two thousand, two hundred people. Four new states joined the union—North and South Dakota, Washington and Montana. Thousands of people, dubbed "Sooners," rushed to claim free land in Oklahoma. The West was beckoning. The Union Pacific Railroad was offering low fares. The rush to the Yukon was in full swing.

At age fifteen, Haskell was sent to school in Massachusetts. His father had hopes that the boy would enter the ministry. Young William, though, had other ideas. Choosing not to go to college, he found work in the retail business field. At twenty-two, he was clerking in a dry goods store and after a particularly trying day began to reflect on his life.

Haskell gave notice to his employer and took a train headed for the sunset. He stopped along the way, earning a little money before moving ever closer to the coast. His feet itching and his adventurous heart yearning, at dinner one night in Colorado Springs, he happened to overhear a conversation at a nearby table. Enjoying dinner were some old-timers who spoke of finding gold nuggets of tremendous value.

Their remarks kindled an interest that turned Haskell's attention to mining. He worked at various places, eventually reaching California. Along the way he gathered experience and knowledge about the elusive yellow metal that men sought, fought over, and sometimes died while chasing. He learned to distinguish truth from exaggerations such as those of the old-timers whose conversation led him to choose his current path.

In 1895, Haskell ran across another young man of about his age. Joseph Meeker had spent a year in Alaska and was fascinated with the place, his fond recollections frequently entering into their conversations. It was not long before the two decided to pool their cash and head for that far-off country. Based on Meeker's experience, they gathered an outfit and set off for Juneau.

Their outfit weighed a bit over three thousand pounds. Hearing the rate being charged by Indian porters—and reports of the unreliability of those porters—they decided to pack their own gear up the dreaded Chilkoot trail. That task eventually completed, they made their way to the lakes where Meeker directed the building of a boat. Haskell almost lost his life navigating the rapids while ferrying a load of their goods.

Belongings of travelers were considered sacred in the North. Haskell and Meeker were outraged when they found a sack of sugar and another of flour missing from their cache after the mishap in the rapids. The items had been left behind during a search for other lost belongings. Sometime later, a suspect was found at a camp downriver. Other travelers were eager to string up the culprit who tried to hide the goods identified by the owners. Haskell, however, could not sanction that sentence. Instead, he administered the fifty lashes agreed to as a minimum punishment. He was relieved to report that the thief turned out to be a good citizen thereafter.

Reaching the Yukon, they began their downstream trek, destined for Circle on the Alaska side of the border. Ironically, they passed with little notice a tributary of the Yukon known as Rabbit Creek and another that had an Indian name ending with what sounded like "dyke." Those creeks would soon become known around the world as Eldorado and Klondike, but the newcomers did not realize that Robert Henderson was prospecting on Rabbit Creek nor had George Carmack yet collected his shotgun shell-full of gold.

The two partners continued on to Birch Creek, on the Alaska side of the border where the Circle claims were found, where they prospected and filed claims. Their mines produced grub money, but did not make them wealthy. They remained there over the winter of 1896-97.

When news of Carmack's discovery reached Fortymile and Circle, both camps emptied. Haskell joined the stampede while Meeker remained behind to work his claim, having seen other rumors turn out to be bogus.

Haskell reached the Klondike and found the stream had been staked to Number Sixty Above—more than five miles upstream of the discovery. As he followed the stream, other hopefuls were rushing past, each looking for a place where they could get in on the newfound pot of gold.

Eventually, Haskell staked a promising claim. Meeker later joined him on the Canadian side and obtained his own plot of ground. They came up with a "comfortable" amount of gold but nothing to match the finds of many others. The two claims, Haskell wrote, were worth about fifty thousand dollars.

The winter of 1896 had been rough and the two were often hungry. When they looked at the prospects for their second winter, they agreed with those who were predicting dire circumstances for the many hundreds of people who were coming in. Boats from St. Michael had not been able to make it all the way to Dawson and some of the last ones in had devoted much of their cargo space to liquor rather than food. Taking a pragmatic view, they decided to head out with dog sleds, by way of Dyea.

It was a fateful decision. As they approached the iced over rivers where Haskell had been dunked in the rapids, Meeker's sled slipped and he was dragged over and into the raging water, then disappeared under a sheet of ice. Haskell made a grab for a rope tied to the sled, only to see it pulled away from his reach. Shocked and saddened, his grief turned into horror when he realized that while he had their provisions and a couple of blankets, the stove and utensils were lost. Apparently also lost was Meeker's gold, although that was not clear.

While Haskell was not one of the miners who hit it rich, he knew many of those who did. His tome lists the mines and their output. He wrote of the traders, confirming that Jack McQuesten was a friend of the miners, advancing credit and grubstaking their prospecting. He gave anecdotes and insight into the lives of Ethel and CJ Berry, the newlyweds who were among the wealthiest of those who took out the "ton of gold" that started the rush.

Haskell's book, while filled with stories of the Yukon, was also one of the most valuable resources for those who were headed to the Klondike.

Unfortunately, little more is known of the man whose recollections so closely pictured life on the creeks and the men and women who poet Robert Service later said "moiled" for gold.

Traders sold supplies, grub-staked early prospectors

Three traders who played important roles in the days preceding and during the Gold Rush—John McQuesten, Arthur Harper, and Al Mayo—were associated with the Alaska Commercial Company. More commonly known by its initials, the AC Company was established in 1868, one year after the purchase of Alaska from Russia. The Alaska assets of the Russian-American Company were purchased by Hutchison, Kohl & Company, then transferred to Louis Sloss, Lewis Gerstle, and A. Wassermann. Going through a series of ownerships, the firm still continues under the Alaska Commercial Company banner and is the largest rural retail chain in the Forty-Ninth State.

McQuesten built several of the AC Company trading posts along the Yukon. His first was Fort Reliance, a few miles downstream from what now is Dawson City. Called "forts," the structures were hastily erected log buildings that served as stores with quarters attached and were not military facilities. In several cases they doubled as post offices and sometimes as locations where miner's courts were held.

McQuesten and his helper, Frank Banfield, were there for the first season. Mayo and Harper were there for the next two seasons, until forced out by threats from the

Natives. McQuesten came back to make amends and bargain with the clan elders. Two elderly women died and a girl had been blinded after eating bread made from flour mixed with grease laced with rat poison. The mixture was found by looters where it had been hidden by Mayo and Harper in the closed trading post. Good relations were restored only after payment of tribute demanded for the injuries. McQuesten was happy that the elders agreed to pay for the items taken from the store, but he only reluctantly added a valuable sled dog to the tribute on which the elders insisted.

Mayo remained at Fort Reliance until 1886. That post, the first established on the Yukon in that area, became Yukon Milepost Zero, from which other locations were measured. Fortymile was forty miles downstream from Reliance. Other settlements became known by their distance from that point.

Fort Nelson opened in 1886 to serve the miners on the Stewart River. Also in that year Fortymile was opened near the mouth of the Fortymile River in the Birch Creek district. When the international boundary was established by United States Army Lieutenant Frederick Schwatka, the North West Mounted Police moved in to enforce Canadian laws and collect customs. McQuesten then moved to Circle City, on the Alaska side, and opened a trading post in that new town that had sprung up after a small stampede.

Initially, the company bartered with residents of the villages along the Yukon, swapping goods for furs, wild game meat, labor, and handiwork. With an eye to increasing business, the traders readily encouraged prospectors to explore nearby streams in their search for valuable minerals. In some cases they "grub-staked" prospectors by supplying them in return for partial ownership of what was found.

Early prospectors trickled in with their packs and pans, sometimes finding colors on the many streams flowing into the great river. When enough colors were found to warrant further investigation, word circulated and other prospectors followed. By the time of the Gold Rush, there were a few hundred people exploring creeks, dividing their time between prospecting and trapping, the latter a more dependable source of income.

Alaska Mercantile Company store at Nome was well-stocked with goods, including cast iron wood-burning range in foreground. Icebound for half the year and freight only coming by sea-going vessels, keeping shelves stocked to serve the populace was a challenge.

Photographer unknown

Jack McQuesten well-liked by prospectors

An early entry into the Yukon history was Captain Leroy Napoleon “Jack” McQuesten, a man revered by the miners he served—and who were to make him rich. He was quoted as telling a biographer that he realized, “If people are busy digging for gold, sell shovels.” It was not his business sense but his heart, however, that earned him the title of “Father of the Yukon.”

McQuesten, standing more than six feet tall, or half a foot taller than the average man of the time, was big even as a child. Born July 9, 1835 in Litchfield, New Hampshire, later moving with his family to Illinois, he was raised on a farm. When he was thirteen, the family was caught up in the California Gold Rush of 1849. While on a voyage with his brother Varnom, McQuesten was credited with saving a schooner and its crew during a severe storm while sailing along the Pacific Coast. The trader’s nickname came from the British term “Jack Tar” applied to sailors; that title solidified many years later when he skippered the first steamboats to navigate the Yukon.

Leaving the sea, the thirty-five-year-old traveled to the Yukon by way of British Columbia, prospecting streams along the Mackenzie and Porcupine rivers until reaching Fort Yukon in 1873. According to his memoir, he and his companions suffered many deprivations during this time. He began working with the Alaska Commercial Company and established a trading post for them at Fort Reliance, near the mouth of the Klondike—very near the future location of Dawson and the site of the discoveries that were to bring throngs of miners. The importance of this post is indicated by the fact that camps such as Fortymile and Sixtymile were named based on their distances, north and south, respectively, as measured from Fort Reliance.

McQuesten formed partnerships with others who were to leave their marks on the North: Arthur Harper, Al Mayo, and Joe Ladue. Each established the trading posts that became so important to inhabitants of the areas. Necessary supplies were carried in stock and were traded for gold or furs brought in by prospectors and trappers. The traders became influential figures in the settlements that sprang up nearby. When matters had to be decided at miner's meetings—the only semblance of law in the place—the traders were usually in charge.

It was here that McQuesten's character came to be trusted. He was known as a man who was honest and fair, completely trustworthy.

That reputation began early and was strengthened by a tragic incident at Fort Reliance. McQuesten had gone to St. Michael for supplies in 1882. Harper and Mayo were running the trading post in the absence of the manager. A local group of Natives were unhappy over the number of people coming to the Klondike after their Chilkat cousins opened the passes that had previously been closely guarded. The traders decided, rather than face the growing animosity, it would be best to make a graceful retreat. They attempted to secure the valuables and tried to hide a mixture of cyanide and grease that was used as rat poison. Some Indians broke into the trading post and took most of the contents. They found the rat poison; not knowing its danger, it was used to make bread. Two elderly women died and a young girl was blinded after eating the bread.

Returning to Fort Reliance and being informed of the situation, McQuesten invited the clan elders to a pow-wow in hopes of restoring good relations with the Natives. He apologized for the unfortunate results of ingesting the rat poison, but insisted that a wrong had been done in stealing the supplies. He billed them for the loss and the elders agreed to pay. Compensation for the two deaths was not asked, but the trader was asked for compensation to pay for the blinding of the girl. He balked at throwing in a valuable sled dog but reluctantly relented.

The Dominion's official surveyor, William Ogilvie, wrote of the incident that "ten skins were demanded for compensation of the injuries to the child. This amount was cheerfully paid and some presents given besides and the prompt payment and kindliness re-established the best of feelings."

The miners' respect was aided largely by the fact that McQuesten extended credit freely. He was quick to "grubstake" a prospector, preceding by many years the "buy now, pay later" theory of business. What was needed in the way of food, ammunition, clothing and tools to get him through a season was issued, simply on a handshake agreement that the prospector would pay up in the spring after washing out gold gathered during the interim. He was even known to offer extensions when full payment was not possible.

An oft-told example of the trader's understanding

nature is the tale of one prospector who came in to settle up a seven-hundred-dollar account, but only had five hundred dollars' worth of gold. McQuesten reportedly told the miner that he could add the deficit to the grubstake for the coming season. When the prospector lamented that he hadn't yet had his "spree," the trader served him from his bar until the spree was over. When the prospector sobered up the next day, the tab was added to the two-hundred-dollar difference and that amount added to the new grubstake. It was generally believed that all his bills were paid in full, as apparently was the case for most of the other miners.

When a competing company set up a trading post at Fortymile, the prospectors showed their loyalty to McQuesten even though the North American Trading and Transportation Company offered lower prices. Captain John Healy's company sent out bills at the end of the month and demanded prompt payment.

In 1894, sixty-nine residents of Fortymile met to form the Yukon Order of Pioneers, whose membership was limited to people there for more than six years. Their motto was "Do as you would be done by." Jack McQuesten was elected as its first president, an indication of the esteem in which he was held by the sourdoughs.

McQuesten's success as a trader was aided considerably by his wife, a Koyukon Athabascan named Satejdenalno. Kate, her Christian name, was fourteen when McQuesten, Mayo and Harper stopped at Kokrine's trading post in 1874, just seven years after the United States' purchase of Russian America. Kokrine was one of a few Russians who remained at the former Russian America Company trading posts. Satejdenalno attended school at Russian Mission and spoke Russian and English as well as her Native tongue. She and McQuesten were married first in a traditional Athabaskan ceremony and later in a Protestant ceremony in 1878 when she reached the age of eighteen. Over the course of several years, they were to have eleven children, eight of them born in a tent outside their house in accordance with traditional Athabascan ways. Superstition held that a birth in the presence of anyone other than a midwife foretold bad fortune, even death.

An intelligent woman who was multi-lingual, Katherine James McQuesten went with her husband as they established trading posts in Circle City and Fortymile. In both places she grew vegetable gardens and led social activities that bridged the cultures of the Native women and the few white women who were in the camps. She served as an interpreter for her husband and kept books, helping to keep track of his many handshake deals.

In 1897, anticipating the shortage of food expected to be experienced by the influx of people flocking to the Klondike, McQuesten moved his family from Fortymile to a Victorian house in Berkley, California. Their children all graduated from high school there. Jack died in 1909 in Seattle while attending the Yukon-Pacific Exposition, Kate eight years later at their home, both having lived out their lives in comfort from the rewards of their time in the North.

Irishman Arthur Harper favored mining over trading

The only one of the renowned trio of traders to be more interested in finding gold than in selling goods, Arthur Harper had gone to California seeking "the stuff of men's dreams" at age 20. Born in Antrim County, Ireland, in 1835, his family had soon afterwards come to the New World.

From California, Harper went to British Columbia during the 1860s, leaving there in 1871 to go farther north. In 1872 and 1873 he prospected along the Liard, McKenzie, Ray, and Porcupine rivers. He arrived in the Yukon in 1874, traveling with Fred Hart, also born in Antrim and the same age as Harper. Their party met McQuesten and Al Mayo at the confluence of the Sikanni Chief and Fort Nelson rivers. Ever an optimist and a prolific letter writer, Harper continually wrote to friends Outside, urging them to join in the Yukon adventure.

With McQuesten and Mayo aboard the *Yukon* when it stopped at Kokrine's while headed upriver from St. Michael, the miner, who was thirty-nine, spotted fourteen-year-old Jennie Bosco in a Native fishing camp near the landing. Following McQuesten's lead, he asked the girl's parents for her hand in marriage. The villagers were having a rough time, with fishing and trapping at an ebb, and an Athabascan wedding was quickly arranged. Jennie was not as well educated as Satejdenalno, but was wise in the ways of subsistence in the hostile environment. She promised to be a good wife and proved so to be. The couple was to have seven children, all but the youngest to be educated in San Francisco—despite their mother's objection.

The youngest child, Walter, was born in 1893. At age twenty, he agreed to lead Episcopal Archdeacon Hudson Stuck's expedition to climb Denali, the mountain later renamed Mt. McKinley and now having its earlier name restored. At the conclusion of the three-month journey into the Interior of Alaska, Walter Harper is recognized as the first person to summit the highest peak in North America.

Despite his preference for prospecting, Harper spent part of his time running the trading posts and helped to build some of them. He was with Mayo at Fort Reliance when the Natives began to rebel against the incursion into their territory by the many gold-seekers who were appearing there. He also played a role in preparing for the rush headed for Fortymile. Based on his long experience, he knew that the rush would be on when word of the discovery of some coarse gold by two Natives McQuesten had grubstaked spread beyond the camp.

McQuesten was in San Francisco ordering supplies for the following season and needed to be warned to increase the order. The river had already frozen over and the only hope was to have a letter hand-carried to Dyea to be sent out on the next steamship. George Williams, who volunteered for the five-hundred-mile trip across frozen tundra, iced-in rivers and lakes and over the daunting Chilkoot Pass, was not a dog musher but a steamboat sailor. He and Indian Bob secured a sled to be pulled by three dogs, one of them never trained to pull a sled. Provisions to last for the expected duration of the trip were packed, along with survival tools. With temperatures dipping to seventy below and a five-day storm forcing them to hole up in a shelter, their food ran out.

Williams became ill and very weak, being carried part of the way by his companion. Both men suffered frostbite and were about to meet their Maker when a Chilkat party spotted them and hauled them to John Healy's trading post at Dyea. Williams died soon after they arrived. Warmed up and fed, Bob knew he had to get the word out. Speaking only broken English, he reached into a bag of beans, pulled out a handful and scattered it across a table. "Gold," he said. "Like this." He pulled out the letter addressed to McQuesten and handed it to Healy. The extra supplies did not reach Fortymile until late in the summer, but starvation was to be avoided over the coming winter.

Harper did not live to see his youngest son's mountain-climbing feat recognized. Coming down with tuberculosis, Harper left the north to seek treatment in the Arizona desert. He passed away in 1897, having gone Outside with the newly minted Klondike Kings on the *Excelsior.* He had transferred his assets to his family and his fare was paid by his employer in recognition of his many years of faithful service. Jennie returned to the Yukon. Walter and his new bride, Frances Wells, were lost in 1918 when the *Princess Sophia* sank in Lynn Canal as they were en route Outside.

Alfred Henry Mayo remained in Alaska rest of life

Of the "Three Musketeers of Yukon Traders," Alfred Mayo was the only one to spend the rest of his life in the north, his Athabaskan wife by his side.

Neehunilthonoth, whose Christian name was Margaret, was the daughter of the chief of Nuklukayet, a village near the junction of the Tanana and Yukon rivers. Her parents were concerned because many of the men who came to the area to take Native women as wives abandoned them. Mayo, a Mason and, as were his trading partners, highly ethical, assured them he would always take care of their daughter, who was only fourteen.

All three—McQuesten, Harper and Mayo—met their brides-to-be, all of them the same age, at Kokrine's when the traders stopped there on the *Yukon* en route from St. Michael to Fortymile in 1884. McQuesten waited until his bride was eighteen before having a Protestant ceremony, while all three were first wed in the Athabaskan tradition. All three marriages were to last "until death did them part." Their unions produced a combined twenty-nine children, Harper fathering seven while the other two had eleven each. Most of those children were sent to schools in San Francisco, headquarters of the trading firm which also operated a steamship line.

Mayo was born February 7, 1847, in Bangor, Maine.

Described as thin and wiry as a youth, he joined a circus and performed as an aerialist and also as a clown. Possessing a dry wit, he loved to play practical jokes. By the age of twenty-four he was in the northland, trapping for furs, trading with the Natives and always prospecting for gold. He traveled at various times with Jack McQuesten, James McKnipp, Arthur Harper, Fred Hart, Joseph Ladue, and Andrew Kansellar.

McQuesten later was to say that Mayo was "the best guy a man would ever want for a trailmate in the Alaska-Yukon region." In May of 1874 they and Harper were at St. Michael at the mouth of the Yukon. McQuesten and Mayo signed on with the Alaska Commercial Company as agents while Harper went with a competing firm—which later folded. On their way upriver they stopped at various villages, one of which was to be where they found the girls they were to marry.

Mayo dropped off at Tanana where he established a trading post. In 1886 the three men built a store at Stewart River, site of the first "rush" to the Klondike, which drew about two hundred fortune-seekers. It was there that Mayo presided over a miner's meeting, a task he was called on to perform many times over the coming years.

Four years later they built the post at Fortymile, site of the next sizeable strike. It was that camp which soon became the district office where Yukon-area claims were recorded. From there, Mayo returned to Tanana in 1894. By that time the three men had become independent traders rather than employees of the AC Company. They bought their supplies from their former employer, but set their own prices and terms in dealing with those who stopped at the stores. In those days there was very little cash and much of the trading was done through bartering of furs, gold or labor. Such an arrangement also allowed the traders to offer lenient credit terms and grubstake prospectors.

Mayo gained the title of Captain after the trio, which operated as an informal partnership, purchased the little steamer *New Racket* from Ed Schieffelin. The Natives joked that the vessel's *New Racket* name was because of the noise made by the clanking of its engine and the sound of hissing steam. Its approach, presence, and departure were always well advertised. Mayo was to skipper the vessel for two decades.

Schieffelin was the wealthy silver miner who founded the town of Tombstone, Arizona. He had a notion that a mineral belt circles the globe and had gone to the Yukon in search of finding quartz veins rich in gold. He and his brother, Al, spent most of 1882 exploring the Yukon, finding small traces of gold, but after experiencing temperatures of fifty degrees below zero, decided that Alaska was not worth further exploration. Schieffelien returned to California without realizing his mistake.

Rampart was home base for the Mayo family from the time the post was established there in 1897. He built the Florence Hotel, named for their daughter. The hotel had rooms, a saloon, and blacksmith shop. Rex Beach and Wyatt Earp both worked for Mayo, according to Mayo's biographers. Beach, in addition to prospecting without success, was a storyteller and actor who took part in performances at various mining camp theaters. Earp was a gambler and saloon-keeper, including ones in Tombstone where he would have known Schieffelin.

Margaret Mayo was a very caring person and served as Rampart's midwife. She also arranged to have the camp's children vaccinated. She died at sixty-five during the influenza epidemic that swept the north in 1925. Her husband predeceased her, passing away in Rampart on July 17, 1925, at the age of 77. Although their children were educated Outside, the Mayos made only one trip beyond the Yukon and its environs, leaving in 1900 and returning a year later.

Steamer *New Racket* tied up at Fortymile trading post.

Photographer unknown

Traders' wives important assistants to husbands

Too little emphasis has been placed here on the role of the wives of the traders. Although only fourteen when they were married, they all were wise in the ways and customs of their people. They quickly adapted to the lives of their husbands. Two of them had been educated in mission schools and spoke English and Russian as well as having learned the basics of reading, writing, history, science, and arithmetic. They were valuable assets in the trading business of their husbands. While McQuesten and the others were extremely liberal in extending credit, it is generally felt that the wives—who were well versed in the Native habits, both of frugality and an eye for profit—played a major role in the success of their husbands.

The women, although quiet and retiring in nature, were extremely helpful in advising their spouses in their relations with the Native trappers who brought meat, fish, and pelts in for trade and those who were employed at the post. As the wives of the village traders, they were looked up to and in turn looked after the people who were customers of their stores. They were highly regarded by the residents, both Natives and whites.

Joseph Ladue's vision makes Dawson founder rich

The founder of the Dawson townsite, Joseph Ladue, had gone north in 1881, one of some fifty people to cross that year into the Yukon by way of the Chilkoot Trail.

Ladue was born July 28, 1855, in Schuyler Falls, New York, the son of a stonemason. His mother died when he was only seven years old, his father passing away twelve years later. Soon after being orphaned, Ladue headed west. Finding himself in Deadwood in the Dakota Territory, he secured a job with the Hidden Treasures Mine. Adept at nineteen years of age, he learned quickly and became superintendent of a sixty-stamp mill. Leaving that job, he prospected in Arizona and New Mexico, hoping to make a strike of his own.

Deadwood at the time Ladue was there was where exciting things were happening. Unrest was high between whites and the Indians and soon George Armstrong Custer would meet his end at Little Big Horn. Wild Bill Hickock would die holding a hand with two pairs—aces over eights—shot in the back of the head by a nobody looking to make a name for himself. The young man may have met Calamity Jane, who would later visit the town Ladue founded but would not meet him there, had by then lost her trim figure and become a heavy drinker, but was still a popular celebrity. Her visit was noted in a couple of lines in the *Klondike Nugget.*

Dawson City was still in the future when Ladue reached the Yukon and met up with Jack McQuesten and Arthur Harper two years after he climbed the Chilkoot. The traders had moved from Fort Reliance to Circle and then to Fortymile when the newcomer joined with them to open a post they called Fort Ogilvie. Ladue set up a sawmill turning out lumber for buildings, sluiceboxes and other items. It was there that McQuesten grubstaked prospector Robert Henderson. When Henderson found gold on the creek he named Gold Bottom, the traders realized that other prospectors would come.

Ladue's thoughts reached beyond the trading post: He saw the potential for a town to spring up. With that in mind, he staked out a township and had streets and adjoining lots surveyed. The lots he offered at prices ranging from five dollars to twenty-five. At first, there was little demand. He filed his township claim on August 28, 1896. Twelve days earlier, George Washington Carmack filed Discovery Claim on Bonanza Creek. Shortly afterwards, Ladue was to sell his lots for prices ranging from eight hundred to eight thousand dollars each. His sawmill had been moved to the town and could not keep up with the demand; he was getting one hundred forty dollars per thousand board feet.

Wealthy beyond his fondest dreams but exhausted under the pressure of the fast growth, Ladue was overwhelmed. He sold the sawmill and made arrangements for someone else to handle the real estate. In 1897, just a year after seeing his vision fulfilled, he pulled up stakes and headed back to New York. He bought a two hundred and fifty-acre farm, married Anna "Kitty" Mason and settled down in his hometown of Schuyler Falls.

In 1900, after two years of wedded bliss, Ladue's doctor told him he should head West, to Colorado Springs where he might relief for the tuberculosis that was consuming his lungs. After but a short time at the resort he decided to return to his birthplace. He died at the age of forty-five, on June 27, 1900, survived by Kitty and an adopted son.

Duped passengers dub Yukon camp 'Suckerville'

The all-water route was used by many of the stampeders who could afford the fare. Steamships from West Coast ports, primarily San Francisco and Seattle, cut across the Gulf of Alaska and stopped for coal at Dutch Harbor in the Aleutian Islands. The passage between Akutan and Unalaska islands still is the primary entrance to the Bering Sea.

The Inside Passage to Lynn Canal in Southeast Alaska is more protected and much smoother. That route was called the "Poor Man's Route." It hugged the coast, sheltered by a line of islands along British Columbia, having only a short stretch of the Gulf of Alaska to cross before entering the protected waterways of Southeast Alaska.

The all-water route was chosen for an expedition led by Seattle Mayor William D. Wood shortly after word of the Klondike strike became known. Wood, a land developer, attorney and president of an electric trolley line, was elected mayor in 1896. Historians in that city chose to remember him for his role in developing the high-end subdivision of Green Lakes, giving only brief mention to his "golden opportunity" in the business of transporting people to the Klondike. Those who record the history of the Gold Rush, however, paint him with a far less friendly brush.

Mayor Wood, then thirty-nine years of age, was in San Francisco attending a Christian Endeavors convention on that July day in 1897 when the *Excelsior* arrived carrying miners and their gold from the north. Among the thirty passengers under the charge of Captain J. F. Higgins were Clarence and Ethel Berry, who alone were reported to have brought with them one hundred thirty thousand dollars in nuggets and dust weighing a little more than six hundred and thirty-seven pounds. Always with an eye out for an opportunity to feather his financial nest, Wood went to the Western Union office and fired off a telegram bidding *adieu* to his post at Seattle City Hall. He quickly organized the Seattle and Yukon Trading Company and raised one hundred and fifty thousand dollars among investors who were eager to cash in on the rush to the Yukon. He chartered the brand-new steamship *Humboldt* and advertised a three hundred dollar fare for the all-water trip, San Francisco to Dawson.

His investors were correct in recognizing a demand for transportation to the gold fields—transportation that seemed to afford a much more comfortable trip than experienced by people taking the "poor man's route" over the passes from Skagway. There were so many people clamoring for passage on the *Humboldt* that Wood sought ways to accommodate more than the vessel was designed to carry. The two-deck, two-masted ship was only one year old, designed for thirty-six crew and one hundred forty passengers. It is not known how many were aboard when she set sail for the Yukon.

After a delay of several days, Wood announced to the crowd waiting at the gangplank that they would not be allowed to carry all the luggage they had brought with them. His plan was to use one of the holds to accommodate bunks for steerage passengers who were willing to pay full price for the chance to sail to the Promised Land. Ever the businessman, Wood had no intention, though, of leaving behind the goods he intended to sell at the store he planned to open in Dawson. In all, he demanded that the ticketholders abandon twenty-five tons of their private possessions.

The passengers were incensed, loudly protesting the idea of leaving behind all but essential belongings. No such restrictions had been placed on their tickets and the anxious crowd insisted that treasured items that would give comfort in the strange land should go with them. There were shouts of "Get a rope," but Wood refused to budge. After all, there were many people at the wharf who were anxious to take the place of anyone who refused to go. That realization finally settled the dispute and the *Humboldt* set sail on August 16, just thirty-three days after the *Excelsior's* arrival.

Thirteen days later, after crossing the Gulf of Alaska and stopping at Dutch Harbor to load coal, the ship anchored off St. Michael at the mouth of the Yukon.

On the banks of the river leading to their destination seventeen hundred miles winding to the north, east, and south, the passengers were hit with another stunning surprise: although their tickets were for passage from San Francisco to Dawson, there was no steamboat waiting for them to board. No arrangement had been made by Wood with the company operating the tiny steamer then on its way upriver in one of the last trips of the season. A boat would have to be built if the fledgling venture was to proceed. It was already the end of August. The

Yukon begins to freeze over in early October. There would be a race against time—and against the odds. It was to be a race won by the weather.

The craft hurriedly put together on the rocky shore of Norton Sound was named by Wood the *Seattle No. 1*. It was called by the disgruntled passengers, though, the *Mukluk* because it very much resembled the footwear worn by the Natives in the area. Another steamer, the *May West,* was also under construction there by another party and a friendly completion competition began. The first to be launched was the *May West,* whose name did not come from the future "Queen of Naughty," who was only three years old at that time, but possibly from Cape West May, a New Jersey borough founded in 1884 by descendants of the *Mayflower* voyagers.

Neither craft made it farther than nine hundred miles, less than halfway to their destination. The *Seattle No.1/Mukluk* had caught up with the *May West* which had left first, only to be blocked by thick ice. Occupants of both steamers camped on the banks of the Yukon; they would not reach Dawson for another nine months. Meanwhile, they pitched tents and erected what shelters they could to wait out the winter where temperatures often hit fifty or sixty degrees below zero. The camp was named Woodworth, a combination of the names of the *Mukluk's* Wood and the *May West's* Captain Worth. To its unhappy residents it would forever remain "Suckerville."

Although the passengers had not yet become miners, frequent miner's meetings were held where residents vented their ire. At one of the first, it was decreed that the goods Wood had planned to sell for high prices at Dawson were to be made available to the stranded travelers—at Seattle prices. The embattled expedition leader, who had attempted to abandon the enterprise at St. Michael but was restrained by the passengers, succeeded in escaping their wrath and a noose by walking back there over the ice. He was said by friendly historians to have sold his transportation interests in 1901, devoting his attention to "continuing his business interests." He died at the age of fifty-nine in Seattle on March 23, 1917, of a "stomach ailment."

The "Mukluk" reached Dawson at three o'clock in the morning of June 25, 1898. Its passengers were hungry, dispirited, their clothing patched with fabric salvaged from flour sacks emptied during their ordeal. Many of the unfortunate travelers had quit at St. Michael and returned on the *Humboldt;* they may have been the more fortunate. Most of those who reached Dawson continued on without pausing, climbing the passes and buying tickets to their homes from Skagway. The handful who remained in Dawson found the creeks already taken up in claims and were left to their own means of survival. Their names did not make it into the annals of the Klondike fortunate and perhaps to their satisfaction are left in anonymity. There is no mention that Wood's investors received any of the doubtful proceeds of the sale of the Seattle and Yukon Transportation Company, or whether any of the passengers received refunds.

The *Seattle Number One* continued to be used for Yukon passage. As for the *Humboldt,* it ended up on the rocks but was resurrected and used in the Alaska trade for some time before being sold in San Francisco for debts and little used by its new owner.

The 'Three Lucky Swedes'

Jafet Lindeberg

John Brynteson

Erik Lindblom

'Three lucky Swedes' made Anvil Creek discovery

The Three "Lucky Swedes" generally credited with making the discovery that started the Alaska portion of the Gold Rush were Jafet Lindeberg, Erik Lindblom and John Brynteson. Lindeberg was actually of Norwegian birth but of Swedish parents. All seeking their fortunes, in 1898 they met by fortuitous chance. Lindeberg came to Alaska as a reindeer herder. Brynteson, an experienced coal and iron miner, had been sent north to look for coal for a Swedish Mission at Golofnin (now Golovin). Lindblom had signed on as a crew member of a ship bound for Kotzebue, but jumped ship when sent ashore to collect fresh water.

In July of 1898, Brynteson joined Rev. Nels Oskar Hulteson and H. L. Blake on a prospecting trip ranging out from Golovin Bay on the southeastern corner of the Seward Peninsula westward to the Bering Sea and north toward Kotzebue. They became stormbound, however, and pulled into shore. Prospecting there, they found some colors but not enough to be worthy of development. Returning to Golovin, Brynteson was to meet Lindeberg and Lindblom. They set out in late August, following the same route and met the same fate, being forced to seek shelter in the mouth of a river. The men picked spots on creeks that flowed into the winding river and began panning in search of gold.

Lindeberg found an exciting prospect on a large creek while the other two also found gold on a creek Lindblom named Mountain Creek and another, Snow Gulch, a short distance away. Lindeberg staked discovery claims on September 20, naming the creek Anvil because of a rock formation that resembled that implement which was perched atop a low mountain from which the river flowed. The river was given the name Snake because of its meandering course. The men staked claims in their own names—and even added some in the names of friends. They hurriedly constructed and set up a crude rocker. A few days later when they left to formalize the claims, they carried with them seventeen thousand dollars in coarse gold pulled from the three streambeds.

Heading back to Golovin, the three prospectors invited J. W. "Gabe" Price, Dr. A. R. Kittleson, and Johan Torneson to join with them in forming the Cape Nome Mining District. Kittleson was named as recorder, the person charged with keeping records of claims that were filed. Lindeberg, Lindblom and Byrnteson also incorporated the Pioneer Mining Company as the organization operating their mining interests. Lindeberg was named president.

The inclusion of Price in the new mining district was a wise move. He had been sent north by Charles D. Lane, a wealthy miner who was interested in rumors of promising mineral finds in the new northern territory. Lane was to found the Wild Goose Company which was to become second only to the Three Swedes' Pioneer Mining in Nome's business community.

This rock atop a mountain at Cape Nome inspired the name for Anvil Creek, site of the discovery that sparked the Nome gold rush. Eric A. Hegg photo
University of Washington Libraries, Special Collections, Hegg 1603

Reindeer herder Lindeberg heads top gold firm

Jafet Lindeberg was born September 12, 1873, in Badderen, Norway, the son of a farmer and fisherman from Norbotten, Sweden. He grew up in an area where iron and copper were mined and frequently accompanied his uncle on prospecting trips. Although gaining only a smattering of experience with mining, he otherwise was well-educated. His teacher made sure he was comfortable in English, French, and German in addition to Norwegian and Swedish.

At the age of twenty-five, he jumped at the chance to sign on with the United States government program to import reindeer to Alaska. There had been extensive advertising as the project officials sought herders who were experienced in handling the animals. Norwegians were expected to teach the Eskimos how to care for the reindeer because they had been shown to communicate better with the Alaska Natives than had been the Siberians initially employed. Lindeberg signed a contract to go to the Bering Sea and sailed on the *Manitoba* on February 4, 1898. He was one of one hundred thirteen passengers, five hundred thirty-five reindeer, and two hundred fifty tons of lichens to feed the animals aboard the ship.

How Lindeberg came to leave the reindeer project is a

matter of dispute. He wrote in a letter to Frank L. Hess that he had been assigned to go to Plover Bay in Siberia to receive reindeer but that the agent there, Captain Kelly, had been driven off by hostile natives. "It was then decided that it would be unwise to send me over there, and, being left without a suitable position, Doctor (Sheldon) Jackson gave me permission to leave the Government employ."

Kittleson, however, stated that Lindeberg had been fired by Jackson. It should be noted, that Kittleson—one of the original stakers in the Anvil Creek area and the recorder in the Cape Nome Mining Precinct—later was an outspoken participant in a miner's meeting and supported the failed attempt to declare that Lindeberg's claims should be voided. Still another account is that Lindeberg broke his contract to go prospecting. Kittleson's version is backed up by descendants of some of Lindeberg's colleagues—Samis, the reindeer herders brought over to teach Alaska's Natives how to handle the reindeer. According to a statement in *Báiki: the International Sami Journal,* "One member of our expedition (Jafet Lindeberg) was the luckiest [of those who turned from reindeer herding to prospector]. He was supposed to travel to Siberia to buy more reindeer, but he refused to go. So our boss Sheldon Jackson got angry with him and he lost his job."

Whichever version is correct, it is without dispute that Lindeberg arrived on the Seward Peninsula and was a discoverer of the Anvil Creek riches. He was named president of Pioneer Mining Company and launched other successful businesses in Nome. With James M. Davidson he built the electric light and power company and with Charles Lane formed the Moonlight Springs Water Works Company. Both were valuable assets. Nome historian

Carrie M. McLain, who went to Nome in 1910 and became a teacher and Nome's historian, wrote that Nome policemen were reluctant to venture past the downtown area illuminated by lights from the businesses; as a result crime in the rest of the darkened town was extremely high. The Water Works not only provided clean water for drinking but installed ditches to bring water on which the hydraulic and dredging mining operations depended.

Lindeberg brought a fisherman from Norway, Leonhard Seppala, to Nome to work for him. Seppala was later to gain acclaim as the dog musher who completed the most hazardous leg of the diphtheria serum run from Nenana to Nome in 1925. The serum was needed to combat a diphtheria epidemic threatening the isolated town.

When Alexander McKenzie brought to Nome the corrupt Judge Arthur Noyes and other henchmen he had helped gain public office, Lindeberg was the primary target of their illegal scheme to take over his company's mines. Lindeberg arranged for the successful appeal to the Ninth Circuit Court of Appeals in San Francisco. While awaiting word from the court, he led an armed vigilante group to rescue gold that had been illegally taken by McKenzie. Author Rex Beach, a resident of Nome during that period, wrote a best-selling novel based on that notorious event. It became a stage play and then a movie that was remade four times. The original version was a silent film. In the second, Lindeberg was portrayed by John Wayne as the fictitious character "Glennister." It was said that the famed Nome miner saw the 1942 version of The Spoilers, but thought the John Wayne representation of him was "not really very close."

Lindeberg continued to manage Pioneer Mining until 1926 when the company was sold to Wendall P. Hammon. A present-day firm, Nome Gold, Inc., traces its beginnings to Lindeberg's company.

After leaving Alaska, the historic figure invested in a number of enterprises. One of those was the Swedish-American Bank headquartered in San Francisco and with branches throughout the West Coast. Lindeberg was president of the bank and Lindblom went there with him, serving as vice president. The bank's failure on January 17, 1921, brought more legal problems—with gunplay involved at one point.

Depositors in Tacoma, Washington, filed suit against the bank's officers, claiming mismanagement and demanding that they cough up four hundred sixty-four thousand, nine hundred and fifty-five dollars. A grand jury in Pierce County indicted Lindeberg and issued a warrant for his arrest. Before it was served, though, he was back at his home in San Francisco. Two peace officers were sent to take him into custody. They reported back that someone at the residence displayed firearms, causing them to return to Tacoma empty-handed. It is not known at this late date if they had read Beach's 1906 book or seen the 1914 original movie, both of which presented Lindeberg as a person not to be imposed upon, as the reason for them deciding to change their minds. An extradition order was issued, but California authorities did not comply with the wishes of their northern neighbors. Lindeberg's attorneys argued that the Washington officers had attempted to kidnap his client and that the Pierce County prosecution was unjust.

Lindeberg died in San Francisco in November of 1962 at the age of eighty-nine.

John Brynteson had prior mining experience

The one with the most mining experience among the three Anvil Creek discoverers was John Brynteson. He had worked for several years in copper, iron and coal mines in Michigan. Born August 13, 1871, in Atemarke, Sweden, he came to the United States at the age of sixteen, migrating to Michigan where he became a citizen of the United States in 1896.

While in Michigan, Brynteson was asked by an official of the Swedish Mission Covenant to assist them in locating coal for a mission in Alaska. The Rev. Nels Oskar Hultberg had established a mission school on Golovin Bay some years before and was seeking a less expensive source of fuel to heat the buildings. Brynteson left for Unalakleet in the spring of 1898.

The search for coal was not successful, however Brynteson found that Hultberg was also interested in checking out rumors of gold being found elsewhere on the Seward Peninsula. In July, the two went with H. L. Blake to prospect on various streams. They found some colors, but not promising enough to stake claims. Brynteson was with Lindeberg and Lindblom when they pulled into the mouth of the same river in mid-September.

Brynteson was the first of the trio to leave Alaska and the only one to return to the country of his birth. He married Emma Forsberg on May 2, 1900, in San Francisco. The couple had three children. They purchased a farm, which later became known as The Pruneyard, in Santa Clara County, California. Brynteson invested in several businesses and gave much to philanthropy. He established a home for sailors in San Francisco and financed charitable endeavors in Chicago. Returning to his birthplace, he financed construction of the Ice Palace which housed events in the 1912 Olympic Games held in Stockholm. He died in Svaneholm, Säffle kommun, Sweden, in 1959. He was eighty-eight.

Erik Lindblom jumps ship, becomes one of discoverers

The oldest of the "Three Lucky Swedes" was Erik Lindblom. He was trained as a tailor but picked up some mining experience in the West after coming to America in 1886.

Lindblom was born June 27, 1857, in Hasledglen, Sweden, the son of a school teacher, Olof P. Lindblom. Upon the father's death while Erik was still a youngster, his mother, Brita, struggled to keep her family alive. At seventeen, Erik traveled to London where he attended the YMCA Polytechnical School. Working as a tailor, he met Mary Ann Smith, whose father was also a tailor.

The young couple married on August 2, 1886, and traveled to America. Erik became the father of a son, Olof Henry, who was born on an Indian reservation in Idaho, and a daughter, named Brita after her grandmother.

Lindblom became an American citizen in 1894 while in Montana. He moved his family to San Francisco in 1893 and was engaged in mining in California and Nevada. When he heard of the strike in the Klondike, he signed on as a sailor on the *Alaska*, a sailing ship headed for St. Michael. The ship was in danger of being iced in while in Grantly Harbor, near Teller, when Lindblom was sent ashore with a party assigned to bring back fresh water. Having reached his goal of getting to Alaska, Lindblom jumped ship and hid while the party gave up looking for him and returned to the ship. A search for the missing seaman was underway when Lindblom met a Native trapper taking a boatload of furs to Golovin. He persuaded the man to give him passage and hid under the pelts as they made their way out to sea, headed south. Lindblom was later to recount that the trip was hazardous and "quite unpleasant."

At Golovin, Lindblom met his future partners Lindeberg and Brenteson, as well as trader John Dexter, H. L. Blake, and missionary Nels Oskar Hulteson. Lindblom was to prospect for Hulteson for some time. In September of 1898 he joined Lindeberg and Brenteson on a prospecting trip. Running into bad weather short of their intended stop farther north, they pulled into the mouth of the Snake River. There they took samples on three different creeks where Lindblom had earlier found some colors. All three streams proved to show values indicating excellent promise. A creek which Lindeberg named Anvil was the richest, while the others were also valuable.

The three went back to Golovin and formed the Cape Nome Mining District. When word of their discovery spread, it sparked a second stampede, this one to a place easier to reach than the Klondike.

Lindeberg, Brenteson and Lindblom formed the Pioneer Mining Company to develop their claims. It was to become the biggest producer of gold on the Seward Peninsula. Lindeberg, the most educated of the trio, was named as president and Lindblom as vice-president. The company was targeted by Alexander McKenzie, a North Dakota Republican moneyman and power broker, who came up with a scheme to challenge their claims as being owned by "foreigners." Actually, Lindblom had been naturalized in 1894 while another also had been naturalized before going to Alaska and a third had filed his intention to become an American citizen.

Lindblom left Alaska after a few short years and moved to Oakland, California. There he bought the Claremont Hotel. He continued to be involved in mining in California and Nevada and ventured into Mexico. In Parral, Chihuahua, Mexico, he owned the Parral Electric, Water and Telephone Company.

Erik and Mary Ann Lindblom were divorced in 1907. He married Hanna Sadie Ulrika Sparman that same year. On a visit to his homeland in 1915, Lindblom was knighted by King Gustavus who bestowed on him the Royal Order of Vasa.

Lindblom died in 1928 at his residence in the Hotel Claremont. The hotel was to continue in ownership by the Lindblom family for another decade.

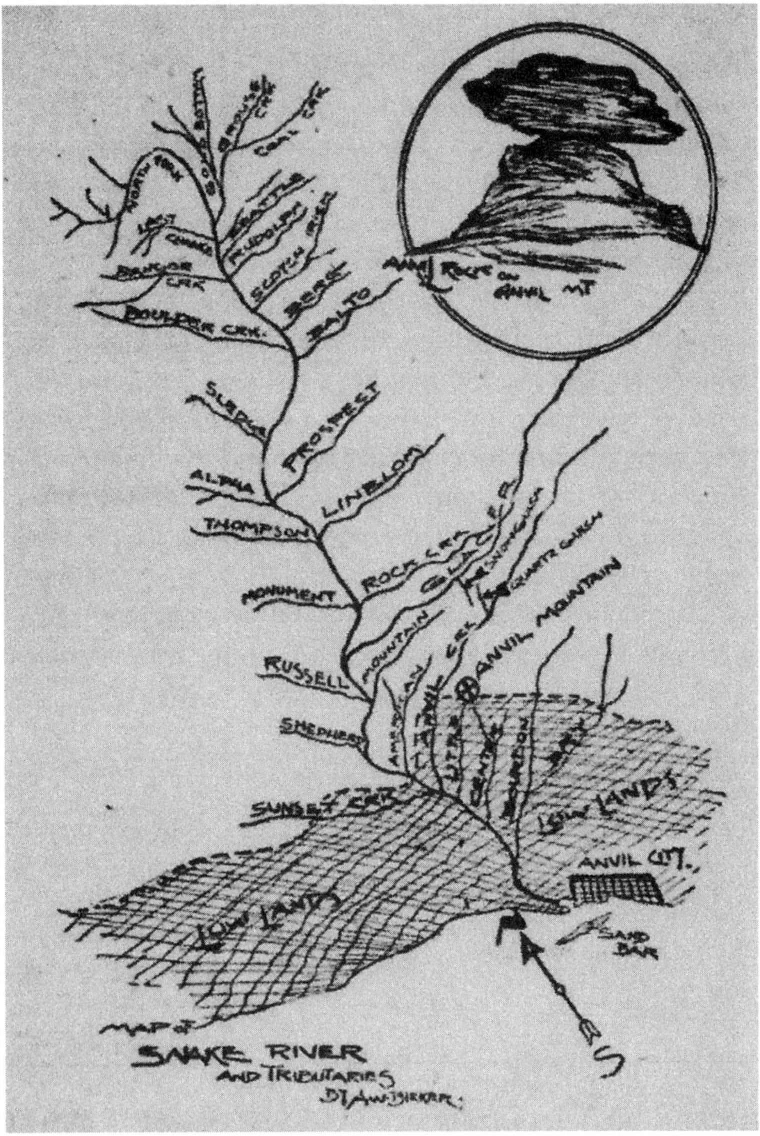

Miner A. W. Bikker mapped Anvil Creek area claims for the Seattle **Post-Intelligencer.** ***At top is rock that inspired the names of the mountain and the creek where Anvil City was established. The site in 1901 became the City of Nome.***

Courtesy Library of Congress Newspaper Archives

Nome miner describes conditions in early Nome

News of the Nome gold strike was of high interest on the part of Seattle residents and the newspapers of that city eagerly interviewed people returning from the north. The Seattle *Post-Intelligencer* in its August 23, 1899, edition prominently displayed a map of the Anvil Creek area above an account given by A. W. Bikker. Originally from Richmond, Virginia, Bikker at the time claimed Cape Nome as his home.

He warned travelers to make the trip out from Nome in the summertime, citing his experience making the journey from St. Michael to Nome, a distance of 200 miles, in the winter. The route is treacherous, he told the reporter. "The most a man can carry is a pack of fifty pounds. A horse cannot pack more than three times that much," he was quoted as saying.

Reporting on the wealth of the Anvil Creek claims owned by the Three Lucky Swedes, Bikker said he saw twenty thousand dollars "washed out of one cleanup on Number One Below on Anvil." He also told of the largest nuggets found, weighing out at eighty-five, seventy, and sixty-one dollars, respectively.

Business in the new camp was brisk, the story read. According to Bikker, the Northern Saloon cost fifty-five hundred dollars to build, was sold three months after it opened for ten thousand, then three months later was sold again for sixteen thousand.

"Gambling is everywhere," Bikker said. He predicted that pastime would grow rapidly because "the men have no other excitement."

California mining mogul becomes major Nome figure

The owner of the second-largest business enterprise in Nome was Charles D. Lane, a California millionaire who had prospered from mining. He formed the Wild Goose Mining Company in Nome and also built the Nome electric utility and the Moonlight Springs Water Works which constructed ditches and wood-stave pipelines to bring water from the mountains to the city and surrounding mining operations. Lane also built a narrow-gauge railroad from Nome to Dexter Creek. The train hauled both freight and passengers. Nome historian Carrie McLain recalled that the line was a boon to the women of Nome and its environs, making it possible for them to comfortably get out of town. Particularly pleasing was that it gave "easy access on nice fall days to the plentiful berry patches" along the route.

Lane was born November 15, 1840, in Pelmyra, Marion County, Missouri. The family moved to take advantage of the California gold rush of 1849 when Charles was twelve. By that winter the youngster was already working in the mines. An industrious lad, he farmed, mined, and drove ox-cart teams. He later said he considered every job a learning experience and entered into each one with enthusiasm.

He went to Idaho where gold had been found on the Snake River and worked in the placer mines there. He developed the Utica Mine at Angels Camp, California, and the Fortuna Mine in Arizona. He had met San Francisco financiers Hayward and Howard who backed his many investments. By 1905 his mines had produced seventeen million dollars in gold. In 1894 he and Hayward consolidated several claims in Juneau, Alaska. They sold those in 1897 after a series of avalanches damaged their operations.

When word of the Klondike discovery reached San Francisco, Lane was in California. Based on information from the Seward Peninsula circulating around mining circles, he sent Gabe Price, who was well versed in mining matters, to Kotzebue to investigate the situation and report back. Price was in Golovin when the three Swedes came in to report their findings and form a mining district. He did not include Lane in the new district but reported back to his boss that he thought highly of the three men. He recommended that Lane support them and he, himself, gave advice to the men who were starting out.

Lane went to Alaska to look things over and formed the Wild Goose Company and the other endeavors. When the claim-jumping scandal arose, he stood with Lindeberg and his partners and actually joined in the armed rescue of claims that had been taken over by the jumpers. He helped finance the appeal to the Ninth Circuit Court of Appeals which returned the mines to the owners.

Lane's wife Anna and their son Tom backed the only successful hard-rock mine on the Seward Peninsula. Their Big Hurrah mine was located east of Nome.

Lane was popular among the miners. When he left the territory to devote his time to his other enterprises, Charles C. Hawley gave a toast at the farewell banquet. Noting Lane's birthplace, he based the ditty on that.

We've all abused Missouri,
And sung our songs of Pike;
And laughed to poke some wicked joke
At raw-boned hungry Ike.
But we've got to pull our houses up,
And 'fess up flat and plain;
Can't find no mate to match the State
That gave us Charley Lane.

Visitors examine cleanup at No. 1 below on Anvil Creek.
University of Washington Libraries, Special Collections, Hegg 1228

Claim-jumping scandal brings Nome lasting notoriety

Before we look closely at the attempt to steal millions in gold from the men who owned claims, let's first consider who and what was involved.

The mastermind of the scheme was one Alexander McKenzie, a political king-maker who was born in Ontario, Canada, but while still a child fled to the United States. His new home had no king, but plenty of elected officials to be manipulated. He was Nome's Soapy Smith—more intelligent, more devious, more refined, but less inclined to put himself in physical danger. He also had a gang, but a smaller one that operated more subtly. As did Soapy's gang, the McKenzie group had some support from among the citizenry who did not unanimously see him to be evil.

As Republican National Committee Chairman for North Dakota for two decades, McKenzie had played a part in electing both of his state's senators and at least one each from Montana and Minnesota. They were in his pocket and owed him big time.

Exactly who were his cohorts among the early miners on Alaska's Seward Peninsula is not clear, although many envied the success of the now rich leaders who had trouble pronouncing words that started with the letter

"J," it coming out as "Yay." Neither is it clear whether McKenzie's plan was sparked by news reports of the discovery by the "Three Lucky Swedes" or was primed by people at Nome who held a grudge. Whichever the case, the scheme was based on the presumption that the most valuable claims were held by foreigners. The McKenzie story line was that their claims could be "jumped" by over-filing. A lawsuit would result. A judge and court officials to be placed there by the McKenzie machine would see to it that McKenzie would be appointed as receiver of the assets until the matter was settled. It would not be settled until McKenzie had arranged to have all the gold taken out and put into his care. He and the gold would then somehow disappear after paying off his henchmen.

The plot, now hatched, had to be quickly put into play. There were legal hoops to be jumped through, but that was nothing new to the man who as a teen-ager had scouted for Colonel George Custer, the ill-fated Indian fighter.

Alaska in those days essentially was lawless. Since the purchase in 1867, it had been under the control of the Army, then the Navy, and finally the Revenue Department. The latter, then in charge, was more interested in collecting customs fees on liquor than on public safety or local government. Alaska's non-Native population was small, the most influential force being missionaries who had gone to the new frontier to save the souls of the "heathens" who had inhabited the area for centuries. Others who went there were mostly adventurers who sought opportunities to exploit the natural resources of the far-off and forbidding land.

Word that gold had been found in some areas in the Yukon River basin had brought a few others whose sparse trappings now included gold pans and shovels. Those already in the north heard of the Klondike discovery soon after George Carmack filed his claim. When the arrivals of the *Excelsior* and the *Portland* with their golden cargo were reported under big bold headlines, the stampede was on. Most of the gold-seekers took the short "poor man's" route to the Klondike via Skagway. A few chose the all-water route with their launching point at St. Michael's, situated at the mouth of the Yukon near Alaska's Seward Peninsula.

Among the prospectors already in Alaska were the three Scandinavians credited with starting the rush to Nome a year after the big invasion at the headwaters of the Yukon. As was the case with the Klondike, claims were filed up and down any creek where flakes of gold had shown up—and even on creeks where the yellow stuff was only in the mind's-eye of the beholder. Not everyone got in on the big payday—and therein lay the rub. Disappointment was high. Jealousy added fuel to the smoldering fire.

Word of the Nome discovery raised concerns in the nation's capital. Congress looked at the Organic Act, the 1884 law that had declared Alaska to be a district. The Act decreed that the laws of Oregon would apply in the new district "so far as the same may be applicable." Among the many provisions was one applying to mining claims. One of those clauses read, "(A)n alien may acquire and hold lands, or interest therein, by purchase, devise, or descent, and he may convey, mortgage, and devise the same, and if he shall die intestate, the same shall descend to his heirs; and in all cases such lands shall be held, conveyed, mortgaged, or devised or shall descend in like manner and with like effect as if such alien were a native citizen of this state or the United States."

Revisions to the Organic Act were made in Senate Bill 3919, introduced on March 1, 1900. While the Act was being considered, Senator T. H. Carter of Montana, not a McKenzie confederate, wanted to make sure that clause and others were retained. North Dakota Senator H. C. Hansbrough, who was in McKenzie's pocket, attempted to strike the alien-applicable clause and instead add one designed to invalidate the Anvil Creek claims, thus opening them to jumping. Rising in opposition to Hansbrough's effort were Senators W. M. Stewart of Nevada, J. C. Spooner of Wisconsin, Knute Nelson of Minnesota, and H. M. Teller of Colorado. Losing support, that amendment was pulled and another substitution was offered by Hansbrough on April 4. That one was more specific, reading, "That persons who are not citizens of the United States or who prior to making location had not legally declared that intention to become such, shall not be permitted to locate, hold, or convey mining claims in said district of Alaska..."

Because more than a month had passed and the new district was desperately in need of a means to provide local government, the bill passed with neither permission nor denial of the right to file by aliens. Overlooked by McKenzie and Hansbrough was the fact that Lindblom had become a naturalized citizen in 1894 and Brynteson two years later in Wisconsin. Lindeberg had applied for citizenship soon after arriving in Alaska in the spring of 1898 under contract to aid in the reindeer project.

One provision of the new law was that Alaska would be divided into three judicial divisions, with a federal judge presiding over the United States District Court in each of the three judicial divisions. Nome was to be in the Second Division. To further the scheme, McKenzie got his cronies in the Senate to persuade President William McKinley to appoint men friendly to the plan to positions in the criminal justice system. McKinley, campaigning for a second term in the White House and anxious to listen to movers and shakers from the Mid-West, accepted the recommendations of the McKenzie men. Arthur H. Noyes of Minnesota was appointed as judge; C. L. Vawter of Montana was named United States Marshal for Alaska; and Joseph K. Wood, also of Montana, was named to the district attorney post. With the chief law enforcement officer to enforce the law, the district attorney to represent the government, and the judge who would make rulings now on board, the conspirators were ready to go.

The three newly appointed officials, two lesser court staffers, and McKenzie left Seattle, ironically aboard the *SS Senator*. Also on board was Robert Chipps of Nome, already primed to cross-file on Lindeberg's Discovery claim on Anvil Creek. Arriving in the Nome roadstead, all members of the party except for Noyes and his wife stepped into the dories waiting to ferry them to shore. The judge was to wait until the next day, giving the others time to get the swindle underway.

With legal papers in hand, Vawter and his deputy the following day made the short trip to Anvil Creek and served the documents on Lindeberg, ordering him to cease work. When the mine owner balked, the marshal went back to the judge at his hotel room. Noyes, without even reading the document put before him, affixed his signature and handed it back to the marshal. At the mine a few minutes later, Lindeberg was told that the mine and its assets were impounded and McKenzie was appointed as receiver to guard the assets until the case had been decided.

Lindeberg complied but immediately contacted an attorney in Nome. An appeal was filed with Noyes but was placed in a pigeonhole without comment.

Unknown to Lindeberg and his attorneys at the time, McKenzie's attorney, W. T. Hume of Hubbard, Beeman, and Hume, were in the process of distributing stock in an Arizona corporation capitalized at fifteen million dollars, with McKenzie holding controlling interest. Half of the anticipated proceeds from the disputed mine were to go to the corporation. Hubbard was given stock in the corporation and in turn McKenzie was given one-fourth interest in the legal firm. Wood, the United States District Attorney, became a silent partner in the firm that represented McKenzie.

Nome's newspapers jumped on the story, whipping up public interest. The editors quickly recognized the potential for expanding their circulation and were eager to gain an advantage in the battle for subscribers. When one took the side of the established miners, another sided with McKenzie and the officials who were his cohorts.

The takeover of the claims of Nome's leading miners gave hope to the unemployed people cursing the "foreigners" who had discovered the gold and located claims early. Two old-timers joined in the clamor to right the "wrongs." One was John Dexter, a Seward Peninsula trader and prospector who had trained Natives in the art of prospecting, staking them to explore creeks in areas where they traveled. He had been left out when the mining district was formed. The other was Rev. Hultberg, whose party had included Brynteson earlier in 1898 when they found what they thought to be poor pay on the creek that became Anvil and was now making Lindeberg rich.

The unhappy group spread word among sympathizers that they were calling a miner's meeting to make some changes. A resolution had been prepared that would void all existing claims and open them to new filing. At the meeting, tempers flared and it was obvious to all that the resolution would have passed. Mine owners who were incensed over the seizure of Lindeberg's claims were determined that they would not stand idly by but would defend their claims at all costs.

How serious was the threat was soon to become clear. Lindeberg got wind of the meeting and learned of a plot that was not public knowledge. A favored group had people posted at various spots, poised to cross-file on the most valuable properties. At the moment the resolution that would void existing claims was passed, a bonfire was to be lighted as a signal for stakes to be driven at the then-to-be-vacated claims.

Armed with this information, Lindeberg contacted General Russell, commander of the newly-created Army garrison at Nome's Fort Davis. Russell dispatched Second Lieutenant Oliver L. Spaulding, Jr., with two soldiers to keep order. One account of the events stated that Spaulding "forced adjournment by motion." Another report, much less restrained in its account, indicates that any motion to adjourn was not of a parliamentary nature but that of bayonets being attached to rifles being carried at the ready.

Realizing that appeals to Noyes would be fruitless, Lindeberg, Lane, and lawyers Samuel Knight and W. H. Metson drafted an appeal to the Ninth Circuit Court of

Appeals, reporting the facts and requesting that Noyes' rulings be overturned. That court was not then in session, but Judge William W. Morrow heard their story and issued an order directing Noyes to make no more decisions regarding jumped claims and ordering McKenzie to return all property taken from the mines. When Noyes ignored the higher court's order, Morrow ordered McKenzie's arrest for contempt of court and sent a stern reprimand to Noyes, ordering him to appear in San Francisco.

At a hearing before the Ninth Circuit Court, McKenzie was found guilty of contempt and given one year in jail. Noyes was fined one thousand dollars. After serving three months, McKenzie was pardoned by President McKinley who was in San Francisco and received the petition for leniency due to McKenzie's "poor health." It was said that the allegedly ill swindler was spry enough to sprint through the door and to the railway station to grab a train back to his North Dakota home once the petition was granted. He was to live for another two decades.

Noyes apparently paid the fine levied against him but did not account for thirty thousand dollars in gold that reportedly was in his possession when he left Nome. Mrs. Noyes remained in Nome for several months after her husband left to face charges.

Judge James Wickersham was called from the Third Division seat at Eagle to handle matters at Nome until a replacement could be appointed. Wickersham dined with Mrs. Noyes and wrote in his diary that she seemed to be completely unaware of her husband's shenanigans and was chagrined to learn that he was disgraced.

Noyes was removed from office by President Theodore Roosevelt, who succeeded McKinley after the latter was assassinated. McKinley was shot on September 6, 1901, by an anarchist who confronted the president at the Pan-American Exposition in Buffalo, New York. One shot was deflected by a button on McKinley's coat but the second projectile was never found. It at first appeared that McKinley would survive, but gangrene in the lining of his stomach felled the nation's leader early in the morning of September 14.

The brains of the failed Nome swindle, McKenzie died June 22, 1922, in St. Paul, Minnesota. He had six children from two marriages. Despite being disgraced in connection with the Alaska affair, he was remembered in his power seat of North Dakota. McKenzie County is named for him, as were the cities of Alexander and McKenzie.

Alexander McKenzie claim-jumping mastermind

Alexander John McKenzie was born in Ontario, Canada, on April 3, 1850. His parents had gone there from Scotland. The boy attended school through fifth grade, then the eleven-year-old left for the United States to seek his fortune.

Little is known of the paths that search took during the period of the American Civil War. However, in 1866 at the age of sixteen, he was reported to have been a scout for Brevet Major General George Armstrong Custer. Early that year, Custer was mustered out of the Union Army as a captain. During that time he considered going into politics. In September, the soldier and his wife accompanied President Andrew Johnson on his railway "Swing Around the Circle" aimed at gaining support for the administration's policy on treating the defeated southern states. Shortly thereafter Custer was reactivated as a captain assigned to the newly-created Seventh Cavalry headquartered at Fort Riley, Kansas. At the behest of General William Sherman, Custer was again promoted to brevet major general. His troops were in the Dakota territory when gold was discovered in the Black Hills.

When the young McKenzie's service as a scout started and ended is unclear. One can only surmise that the teenager may have listened to Custer's stories of political influence and that the general's disdain for any authority other than his own may have rubbed off on the young man. There definitely are some parallels between the two men's careers.

During the period 1874 to 1886, McKenzie served as sheriff of Burleigh County, North Dakota, and later was appointed as a Deputy U.S. Marshal for North Dakota and its southern neighbor. When North Dakota gained statehood in 1889, McKenzie immediately became involved in politics. Seeing that his fortunes lay better in getting people elected to office rather than serving, his tentacles spread throughout the new state. He became the Republican Party chairman and served on the National Republican Committee as a representative from his state for twenty-one years. That post he resigned in 1908, two years after his machine was voted out of office.

Rocking on the Nome beach proved profitable for many.

University of Washington Libraries, Special Collections, HEG 1355

Judge James Wickersham sent to Nome to bring justice

Sent to clean up the judicial mess at Nome left by Arthur Noyes was James Wickersham, judge of the Third Judicial Division based at Eagle. Both men had been appointed to their Alaska posts by President William McKinley in 1900. In contrast to the corrupt previous occupant of that bench, Wickersham was a hard-working man who was truly judicial in his decision-making.

Wickersham was born August 24, 1857, near Patoka, Illinois. After graduating from the eighth grade, he taught in a rural one-room school. He began working as an office boy in the law office of John McAnley Palmer where he read the law while continuing to teach school. He roomed at the home of Isaac Bell, the local postmaster, whose daughter Deborah soon caught his eye.

In January of 1880 the young lawyer passed the Illinois bar exam, obtained a position with the United States Census Bureau and on October 27 married his landlord's daughter. Three years later the couple moved to Tacoma, Washington, where Wickersham first worked as a carpenter. In the fall of 1883 he formed a partnership with Ezra Meeker and both became charter members of the Tacoma Bar Association.

Although in Alaska never to be accused of wavering

in his beliefs, Wickersham experimented in politics. He ran for county probate judge in Tacoma in the fall of 1884 and won a pair of two-year terms. He was to join a labor union party, a businessman's party, the Democratic Party and then the Republican Party. He also was to have three different partners in a space of that many years. His vacillations during his political apprenticeship led to an incident that continued to plague him throughout his life. A pretty young woman named Sadie Brantner accused Wickersham of seducing her. He was tried and found guilty and immediately filed an appeal, alleging misconduct, saying that political opponents, and not he, had played a part in the affair. The appeal consideration dragged on with the press very much interested in the sordid details. While Wickersham stuck to his claim of innocence, Sadie was quite willing to talk. Fortunately for the accused man, she couldn't keep her stories straight. The case was dismissed due to her obvious lack of credibility.

The Wickershams moved to Gig Harbor, Washington, where he practiced law and also developed a residential subdivision. In 1889 he ran as a Republican for the Washington State House of Representatives and won a two-year term. With President William McKinley considering a run for re-election to a second term the following year, gaining favor in the new state in the northwest corner of the country was of interest to the campaign. Just as McKenzie was pulling strings in Minnesota and North Dakota, Washington king-makers were pushing for posts for their favorite sons. Wickersham was asked whether he wanted to be appointed as Counsel General to Japan or to a newly-created federal judgeship in Alaska. The freshman Congressman said he would be happy with either, but his preference was for the Japanese position. As it turned out, the powers that be gave that spot to a former Wickersham partner—ironically who was suspected of being involved in the discredited seduction matter.

Named as judge in the new Third Judicial District, Wickersham left Tacoma for that far north outpost. He was forty-three, his wife thirty-seven at the time. They had three sons, but only one survived past childhood. The older son, Darrell, had been accepted for appointment to the Naval Academy in Maryland. At Seattle, Wickersham had a brief meeting with Noyes, as directed by his superiors in Washington, to establish boundaries between the court's divisions. Proceeding north, the couple landed at Skagway where they met with the First Division judge for the same purpose. They took the new White Pass and Yukon Railroad to Lake Bennett and from there went by sternwheeler to Eagle, the seat of Wickersham's jurisdiction. That mining town, until an Army survey later confirmed by the Canadian surveyor William Ogilvie, was originally believed to be on the Canadian side of the border. It had once been a thriving mining camp but had dwindled after the Klondike strike.

With Wickersham's arrival, the Third Judicial Division had a court, but no courthouse, no furniture, no jail, and very little money. Wickersham had been given less than two thousand dollars to pay for startup costs. He not only had to build a courthouse but provide living quarters for himself, a clerk and a bailiff. His experience as a carpenter when first arriving in Tacoma came in handy. It was up to the court to sustain itself on permit fees collected from saloon owners, fines, and various other assessments allowed within his jurisdiction. The new judge's work was cut out for him over his first year. Not a minor consideration was the fact that winter was fast approaching.

Winter was not too much of a problem for the man who had preferred being housed in a nice warm building in Tokyo. He thrived in the outdoors and loved exploring the area that now was home. He was curious about his surroundings and made meticulous notes in the diary he maintained faithfully. Those diaries now comprise forty-seven volumes in the archives of the University of Alaska and are an invaluable resource for historians.

Once court was set up, Wickersham oversaw a variety of cases involving everything from thievery to mayhem. His first murder trial was cut and dried and the defendant was found guilty and sentenced to be hung. The most heart-rending for the judge were those cases held to determine whether a person was insane. In the time of the Gold Rush, there were no psychiatric facilities and no provisions for housing or treating them. If a person seemed to not be in his or her right mind and might be a danger to him or herself or someone else, it was obvious that something should be done. The practice then was to lodge a misdemeanor complaint against them, causing an arrest. At trial, a doctor could issue a medical opinion as to the person's sanity. If found guilty of insanity, the convict could be sent to Morningside Hospital in Portland, Oregon.

Wickersham's diary reveals his agonizing debate over the question. He recalled two cases. In both cases, the persons railed against the government and their complaints grew more frequent and more intense. One man along the Yukon complained that a neighboring family repeatedly broke into his cabin and stole items. He wrote to officials, saying he wanted the whole family arrested because he was unable to identify the actual culprit. The response in this case was that the man would have to go to St. Michael to file a report. The victim said he did not have the money for transportation and it was too far to walk and demanded that officials come to his location and make the arrest. In the other case, a doctor became increasingly vocal in making complaints about other individuals in Nome. In both cases, the individuals were charged with minor crimes and at trial witnesses gave testimony pro and con. The man who wanted his neighbors arrested was found not guilty and let go. The doctor was found guilty of insanity and sent under guard to Portland's Morningside Hospital. The judge agonized over a situation that treated the mentally ill as criminals but over which he was powerless to correct.

The frequency and seriousness of the problem is illustrated by a wry oft-repeated joke that circulated in Alaska from 1900 all through Territorial days. It was, "There are three sides to Alaska: Inside, Outside, and Morningside."

Wickersham's jurisdiction covered a huge portion of Alaska. In winter, he made the rounds by dogsled. In the middle of the winter of 1900, he had to go to Rampart to settle an argument over claim-jumping. He and his clerk traveled by sled for a thousand miles over a forty-five-day period. Their expenses were seven hundred five dollars, a sum for which the judge noted in his diary he "hoped for reimbursement." Due to the distance and conditions of that time, pay checks were three months or more late. An athletic individual, Wickersham handled the conditions well even in his first year. He liked to explore and is remembered as the first white man to attempt to reach the summit of Mount McKinley. Called "Denali" by the Alaska Natives and "Bolshaya Gora" by the Russians ("The Great One" and "Big Mountain," respectively), there have been frequent drives to officially change its name to Denali. The change finally was approved in 2015. The McKinley name was given in 1896 by a prospector who wanted to bring awareness that William McKinley was his candidate of choice as President of the United States. McKinley was assassinated, shot and fatally wounded by an anarchist on September 6, 1901, dying eight days later of gangrene.

Ordered to Nome following Noyes' departure for San Francisco and his subsequent removal from office, Wickersham began clearing more than two hundred lawsuits awaiting placement on the docket. He held court daily, often going late into the night. Most of the cases involved claim jumping and their resolution was made simpler by the Ninth Circuit ruling that resulted from the Lindeberg-Lane appeals. The Court ruled that aliens could own land in Alaska, making moot that particular argument as a basis for action against an owner.

Wickersham returned to Eagle upon the appointment of Noyes' successor. After an Italian immigrant, Felice Pedroni, discovered gold near Barnett's Cache on the Tanana River, the Third Division seat was moved to the new town of Fairbanks.

Wickersham resigned as judge in January of 1908 when he ran for Congress as the territory's non-voting Delegate. He served three successive two-year terms until replaced by Charles A. Sulzer, whose election Wickersham successfully challenged in court. He again had to challenge Sulzer in court and served another short term. Wickersham's election campaigns continued to be smeared by the alleged Tacoma scandal of which he had been cleared. His political opponents would not let this bit of gossip go, nor were they reluctant to voice other complaints. In Tacoma, Wickersham had been a member of an anti-Chinese migrant group. In Alaska, He took advantage of opportunities to profit from land dealings in Fairbanks, purchasing several downtown lots. He also filed on mining claims for himself and for his associates. Despite the allegations of wrongdoing, he was re-elected as a delegate to Congress.

During his terms in Congress, where a Delegate has no vote but is given a voice on the floor of the House, Wickersham was credited with many acts beneficial to Alaska. He was successful in having a university established in 1917, the Alaska Agricultural College and School of Mines opening for classes at Fairbanks in 1922. He offered the first Alaska Statehood Bill in support of giving local control over its governmental affairs—something that was not to happen until four decades later.

After leaving Congress, Wichersham moved to Juneau and practiced law there. His first wife, Deborah, who had been in poor health for most of their married life, died in 1926. He married Grace Vrooman Bishop in 1928. Their Juneau home is maintained as a historical site.

Murderer Fred Hardy first legally hung in Nome

The first legal hanging in Nome occurred on November 19, 1902, when one Fred Hardy, who also was known by his step-father's last name, Watkyns, met his Maker. The twenty-six-year-old was hung from gallows installed in an addition tacked onto the Nome ice house across from the jail.

Hardy was convicted by a jury for murdering three prospectors on Unimak Island in the aftermath of robbing the men's possessions. A fourth man was wounded but survived. As presented at trial, Hardy and George Aston removed one load of goods from a tent, including weapons and ammunition. They returned and removed the remainder of the contents of the tent, then tore it down. The prospectors returned just as the thieves left. Owen Jackson, the one who survived, testified that a man, thought to be Aston, was hiding in tall grass, rose up and fired at the prospectors who had just pulled up in a boat, hitting Florence Sullivan, who clutched his knee and fell onto the rocky beach and was later found dead. The other three tried to get back into their boat but another shot from cliffs above the beach struck P. J. Rooney, who "screamed and fell into the surf," Jackson testified at trial. Con Sullivan, brother of the first victim, and Jackson ran toward shelter at the bottom of the cliff, only to be shot. Mortally wounded, Con Sullivan fell and died where he lay while Jackson stumbled to shelter in rocks below the cliff.

Jackson wandered for several days until found by a friendly hunter and taken to an Aleut fishing village. He related that during that time he was terrified that their assailant would track him down and finish the job.

The crime was reported to authorities and a deputy marshal found Hardy and Aston at another village on the island where the two suspects had apparently terrified the Natives. When the deputy confronted the two, he found in Hardy's possession money and a watch that had the inscription, "Julius to Florence." Taken into custody, both men claimed innocence, but Aston was to blame Hardy and testified against him before a grand jury. Hardy was indicted and brought to trial before Judge James Wickersham. He was convicted and sentenced to hang.

Wickersham's diary describes the trial and his reaction: "It was most an outrageous and cold blooded murder [and] the perpetrator ought to suffer death." If the jury convicted Hardy, the judge said he would sentence him to hang, "although there is one thing that I want to see investigated." He did not elaborate, but did impose the sentence.

An appeal resulted in a second trial, with the outcome the same. Judge Arthur Moore, newly assigned as judge in the Second Division, reaffirmed the sentence.

News accounts and Wickersham's diary refer to Hardy's insistence, even at the gallows, that he was innocent. Observers also noted that the condemned man during the trial and later showed no remorse and was defiant. It was disclosed that Hardy with a partner named Williams had deserted their posts on a ship and hid out on one island in the Aleutian chain. Hardy was alleged to have killed Williams and stolen his possessions.

Unimak Island is in the eastern end of the Aleutian Chain, situated east and adjacent to Unalaska Island where Dutch Harbor is the passageway to the Bering Sea from the North Pacific Ocean.

In an unusual twist, Hardy bequeathed his possessions to a Red Cross worker who visited him in jail at Nome and showed sympathy toward the condemned man. In his will written shortly before the sentence was carried out, Hardy stated that Mrs. Mary E. Hart of Nome was to be given his property. The will listed real property consisting of three hundred seventy-five acres of land in Black County, Michigan; one thousand, five hundred and sixty-five acres of land in Lawrence County, Tennessee; and an unknown amount of money on deposit in the First National Bank, Logan, Ohio, left to him by his grandmother, Meg Watkyns. Also bequeathed was a trunk containing a sword, various personal items worth about five hundred dollars and photographs. The photographs were to be sent to Hardy's mother in Tennessee.

Tombstone, Arizona, friends Wyatt Earp, left, and John Clum reunited at Nome where Earp was co-owner of Dexter Hotel and Saloon and Clum was Postal Inspector for Alaska. In Tombstone, Clum was publisher of the *Epitaph* newspaper. Photo courtesy Library of Congress

John Clum establishes post offices in Alaska

It is no surprise that the Gold Rush attracted a wide range of adventurers and that among them were people who were well known. One such was the man sent to inspect and report on as well as establish and equip post offices in Alaska.

John Phillip Clum was from a well-to-do family in Claverack, New York, where he attended school and enrolled at Rutgers University. His schooling was interrupted by illness and he did not graduate. At age twenty-three and prematurely balding, in 1874 he was named an Indian Agent and sent to take charge of the San Carlos Apache reservation in Arizona. The Indians promptly gave him an Apache name that translated as "Boss with the High Forehead."

The young easterner promptly set about to improve the lot of the Native Americans, forming a police force to maintain law and order. The renegade Geronimo was causing trouble and was being vigorously sought by the Army. With his aim centered on protecting his charges, Clum frequently was at odds with the military—who he thought was out to eliminate the Indians entirely.

Clum hoped to take Geronimo into custody in a way that would spare the leader's life and devised a ruse that would accomplish that. A message was sent that lured the chief to a warehouse where the reservation police were waiting. It was successful and Geronimo was captured without bloodshed. Geronimo was turned over to federal authorities but later escaped.

The San Carlos reservation was well run but the agent felt his accomplishments were not being recognized. Other reservations were being closed and their charges transferred to Clum's care. Agents of the closed reservations were no longer needed and laid off or transferred. Clum complained that his workload was being increased, yet he was not being paid more while the agency was pocketing the money saved by the reductions in personnel.

Clum resigned his position but his distaste for bureaucracy was not forgotten. It would soon make itself known through the editorial pen.

In 1876, Clum traveled to Ohio to marry Mary Dennison Ware. On the trip, he took along twenty-two San Carlos men. To help with expenses, Clum stopped along the way to put on "Wild West" shows. Returning to Arizona with his bride, Clum bought the *Arizona Citizen* newspaper in Tucson. His low opinion of the government was expressed on his editorial page where he wrote such phrases as "political double-crossers in Washington."

Heading to the new silver boomtown of Tombstone, he founded a newspaper he dubbed the *Epitaph,* saying, "'Every Tombstone needs an epitaph." His stated intent was to slow the wide-open pace of the frontier town. That stance drew support from a Law and Order group that backed a no-nonsense bunch of town marshals—including the Earp brothers.

Under a new town charter approved in 1881, Clum

was elected mayor by voters anxious to control lawlessness. Tombstone is located not far from the Mexican border and rustlers, smugglers and others bent on criminal activities were frequent visitors. The settlement had been plagued by such miscreants, in particular a loose band of rustlers known as the Cowboys. On October 26, 1881, the Earps and John Henry "Doc" Holliday had a showdown with some leaders of the outlaw band, killing two of them.

Clum's wife, Mary, died after giving birth to a baby girl, Elizabeth, on December 18, 1880, followed shortly by the death of the baby. Clum sold the newspaper and left Tombstone two years later.

Apparently having softened his views about the government, Clum began working with the Post Office Department as a postal inspector. In 1898 he was dispatched to Alaska as a special commissioner "to examine into postal affairs." He was to establish new post offices as needed and equip existing ones.

Clum arrived in Skagway on March 26, inspecting the existing post offices there and at Dyea. He was at the latter place on Palm Sunday, April 3, when a series of avalanches struck the "Golden Stairs" leading to the Chilkoot Pass. Clum pitched in to help recover the bodies of about seventy people who died when heavy, wet snow slid from the slopes and covered the helpless climbers.

Pausing the next day, he established a post office at Sheep Camp, a spot on the trail toward the Pass leading to the border with Canada.

Proceeding to Dawson City to continue his voyage downriver to Nome, Clum there met Ellen "Nellie" Cashman, widely known as the "Angel of the Cassiar" after leading a team through severe storm conditions to rescue stranded miners at that British Columbia location. Cashman and Clum were friends in Tombstone where she ran one of the town's finer restaurants. She had been a member of the Law and Order League.

A post office at Nome was opened in June of 1899, with Joseph Wright as postmaster.

In Alaska, Clum traveled more than eight thousand miles, visiting settlements. Where there was a need for a post office, Clum found someone to be named as postmaster and issued the supplies needed to operate the facility. Where there already was a post office, he provided supplies needed and inspected their records and condition of the facility. He was to establish seven new post offices.

In January of 1906, Clum was named postmaster at the new mining camp of Fairbanks in Alaska's interior. He served there until 1909. He later was hired by the Southern Pacific Railroad and traveled widely, giving speeches to promote the railroad and tourism.

In 1928 he moved to Los Angeles, remaining there until his death in 1932 at the age of eighty.

Disappointed miners start Nome's free mail delivery

Soon after Postal Inspector John Clum arrived in Nome, the postmaster, Joseph Wright, left for a short trip Outside. Clum was filling in for him when a clerk advised that two newcomers wished to speak with him.

The two were Fred Lockley and Ben Taylor, from Salem, Oregon. Lockley, who was to write a book about his six months in Alaska during the Gold Rush, explained that they were postal workers on leave. They had come north hoping to join those getting rich but had found no land to claim and no job to be had. Instead, Lockley advised that he had an idea that he thought worth considering.

While waiting in the long line at the post office's general delivery door, he saw that people stood for hours awaiting their turn. After all that time, there may or may not be a letter for them. To conserve the clerk's time, a person was only allowed to ask for mail for one person. If he was there on behalf of a partner, he had to go to the back of the line and again await his turn.

Lockley proposed that he and his companion be hired to provide free delivery to businesses within the downtown area, which consisted of only a few blocks.

Clum told them to return later that evening. When

they did, they found a typewritten note advising that free mail delivery to businesses was to be put in effect and that Lockley and Taylor were to be carriers.

The two returned to their tent and began planning. They had brought old postal uniforms to use while digging for gold and pulled these out, brushing them off for use the next morning. At daybreak, they headed out to survey the street, making note of the businesses on each side of the road. Lockley took one side and Taylor the other. They quickly determined that everyone would want to get their mail delivered rather than stand in line; the pair quickly devised a plan. Only the business owners and employees would be entitled to receive delivery, thus limiting the "in care of" extra load.

When the businesses opened the next morning, the two visited each and took down the names of the people entitled to delivery. Most were skeptical until shown Clum's typewritten note and assured that there would be no charge. Neither man drank nor smoked; they turned down ready offers of free beers or cigars in appreciation.

Free delivery began the very next day. The two men showed up at the post office at 4 a.m. to begin sorting mail, looking for names that were on their lists.

Lockley explained that Nome had the largest general delivery service in the entire postal service. There were seventy boxes, he wrote, just to hold mail addressed to persons whose last names started with the letter B. The name most frequently found was Johnson. The boxes for Johnsons had five separations just for the first names: A-J, J-M, M-P, P-S, and S-Z.

The two men quickly encountered problems that had not been foreseen. At the time, there were three Peter Petersons in Nome. Letters addressed to that name were left at the post office and people asking for that name and who had no middle name, were called upon to say who they thought the letter might be from and from what city.

Making it difficult for the two carriers was the transiency of the town. Even the businesses were on the move. Lockley explained that the Red Cross Pharmacy opened in a flimsy wooden frame covered by a canvas on Taylor's side of the street. Soon after, Taylor found the business no longer there. Lockley the next day saw it on a lot on his side of the street. Both men revised their lists. The pharmacy moved twice more before it went out of business, requiring more changes to the delivery lists.

A business located in a tent could move to different lots within a short period of time. One man opened a store, sold it for a profit, opened another and sold it, again for a profit—all in a forty-eight hour period.

During the summer, one or more ships were arriving at Nome every day, each carrying mail. That meant the clerks at the post office—twenty-two on each of two shifts—worked long hours sorting mail. On the days when the ships came in, Lockley and Taylor found themselves working from 4 a.m. until midnight.

They asked the Salem postmaster for and were granted a ninety-day extension of their initial leave of absence. At its end they returned to Oregon and resumed their careers. They had not made a fortune in gold, but had a trunkful of memories and the satisfaction of accomplishing something of value to Nome and its population.

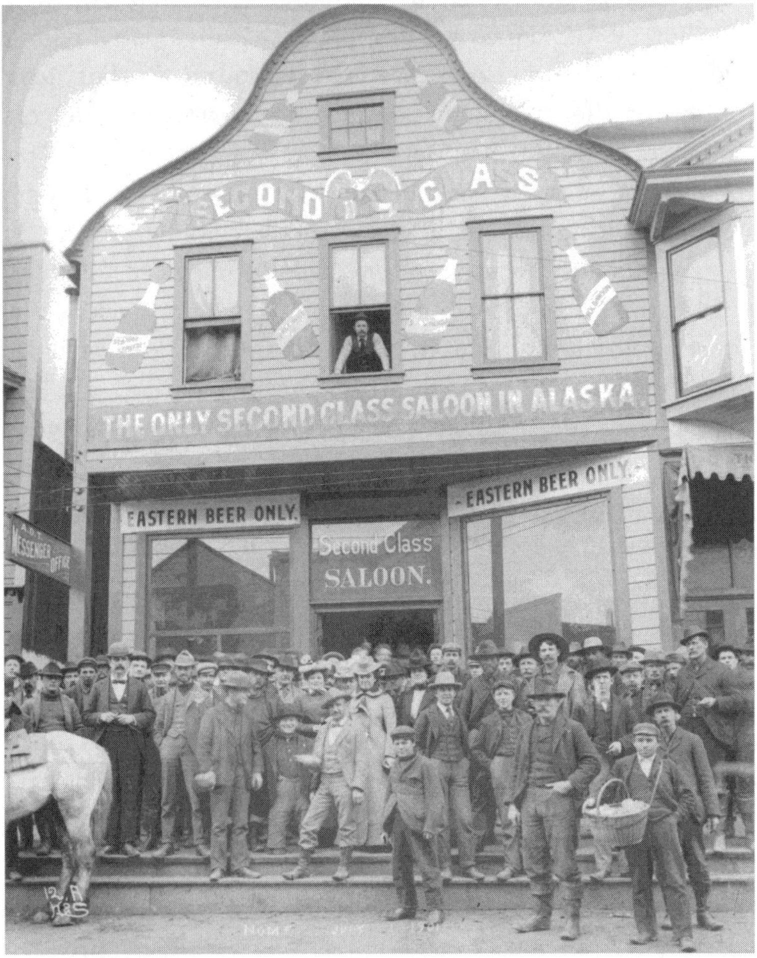

Wyatt Earp owned Nome's Second Class Saloon, a popular watering hole for thirsty miners. Photographer Eric A. Hegg
University of Washington Libraries, Special Collections, Hegg 3019

Wild West legend Earp becomes wealthy hotel owner

Wyatt Earp, now the most legendary gunfighter of the Old West, when he arrived in Alaska in 1898 was still smarting from nationwide criticism of his controversial ruling two years earlier that decided a heavyweight boxing match. Fans—including spectators at the December 2, 1896, San Francisco match between Bob Fitzsimmons and Tom Sharkey—had expected Fitzsimmons to win. Witnesses universally declared that the favorite dominated the bout for eight rounds.

Earp, a last-minute substitute referee, surprised everyone when he ruled that Sharkey was hit below the belt. Fitzsimmons had delivered his famed "solar plexus punch," an uppercut to the midsection that momentarily rendered an opponent helpless. Upon being struck, Sharkey grabbed his groin and sank to his knees. Earp waved Fitzsimmons to his corner and raised Sharkey's hands to a loud chorus of boos.

The loser filed a lawsuit, hoping to overturn the verdict. The San Francisco court, however, after a lengthy hearing decided that because prizefighting was illegal in that city, there were no grounds for action and would not rule as to which man was the rightful winner. That only added fuel to the fire.

Until then, Earp was little known outside the boomtowns of the West. The confrontation of 1881 between Wyatt, Virgil, and Morgan Earp and fellow lawman Johnny Behan and Wyatt's friend John Henry "Doc" Holliday with the "Outlaws" gang in Tombstone, Arizona, was but another gunfight until glamorized later in magazines. "The Gunfight at the OK Corral" brought fame to Wyatt Earp—much to his consternation. It led to the murder of his brother Morgan and crippling of Virgil and led to Wyatt's long odyssey in search of revenge.

Born March 19, 1848, in Monmouth, Illinois, Wyatt Stapp Earp took any job available—often involving a firearm. He hunted buffalo, drove horse-drawn wagons, swept floors, and even occasionally wore a badge.

At the age of twenty-two and having spent several years in Iowa, he married Urilla Sutherland in 1870 and made a home with her. She soon became pregnant, but contracted typhoid fever and died with her unborn baby. Devastated, Earp returned to Illinois. His pursuits in questionable and often illegal activities caused the *Peoria Daily National Democrat* to refer to him after a September 10, 1872, arrest as the "Peoria Bummer."

Earp continued a wanderlust career in search of excitement and riches. He found plenty of the former and little of the latter until he came North in the Gold Rush.

While working as a deputy marshal in Wichita, Kansas, in 1876, Earp decided to follow his brothers to Dodge City, Kansas, where they were looking into the latest booms following discoveries of silver and gold. Two years later he went to Texas where he took a job as assistant city marshal. There, while bursting through the entrance of a saloon after hearing gunshots, he found himself staring into the barrels of six-shooters held by half a dozen liquored-up hoodlums. Suddenly, the leader of the threatening band told his confederates to drop their weapons. Earp was relieved to see someone standing behind the leader, holding a pistol to the man's temple.

Earp's benefactor turned out to be a dentist who had come West in search of a cure for the tuberculosis that was consuming his body. Trading forceps and drill for playing cards, "Doc" Holliday now plied the trade of a gambler. The two men thereafter became lasting friends, bringing the former dentist to join in the shootout that was to become forever famous.

After his brother's murder, Earp began an odyssey to hunt down the members of the Outlaws believed to have been responsible. Two men were killed in revenge. Moving on, Earp followed the trail of new boomtowns, ranging from San Diego, California, to Couer d'Alene, Idaho.

In Tombstone, Earp's business ventures had included saloons and gambling houses. Many of those establishments involved "social" rooms where prostitutes took customers from the bar to enjoy their pleasures. Earp took one of them, Celia Anne "Mattie" Blaylock, to live with him as his common law wife. Mattie died at the age of forty of opium poisoning, officially ruled a suicide.

Earp had already found another favorite among his stable—a comely woman named Josephine Sarah Marcus. She had expected to marry another lawman, Johnny Behan, but he reneged and she went to California. Earp followed her there and she became his common-law wife, called fondly by him as Sadie.

On August 5, 1897, Earp and Sadie departed San Francisco aboard the *Rosalie* for Dawson where Earp and a partner were to establish a gambling house. The couple arrived in Dawson on September 12 after crossing over the Chilkoot Trail. They apparently returned immediately to San Francisco because they are reported as heading back north the next spring via Wrangell where Earp was offered a job as marshal, possibly working there for several days before continuing to Nome.

The Earps got as far as Rampart before being caught in ice when the Yukon froze over. They rented a cabin from Rex Beach, paying $100 per month. After breakup, they continued downriver, stopping at St. Michael. Earp reportedly was hired as manager of a small retail store there before moving on to Nome.

At Nome in the summer of 1898, Earp met Charles E. Hoxie, with whom he formed a partnership to go into the saloon business. They built the Dexter Saloon and Hotel, Nome's first two-story frame building. It measured thirty feet across the front and seventy feet deep, with twelve-foot ceilings and was the town's largest building at the time. Twelve luxurious rooms were located upstairs.

Earp also owned the Second Class Saloon, an exclusive purveyor of eastern beers. There were rooms upstairs and gambling tables on the main floor.

While Earp, by then aged fifty and no longer a spry gunfighter, attempted to maintain the image of a respectable businessman, his past experience in the saloon business could not be shaken. The rooms upstairs in the hotel were frequented by miners anxious for companionship—which the girls in Earp's employ were happy to supply. Earp, who always carried a pistol, whiled away his time at the card tables on the main floor.

Sadie was known in Nome as Mrs. Earp and was accepted in social circles. The Earps were considered to be good citizens in the new frontier city, supporting improvements and law and order. Records of the Nome School

Board in 1900 show a contribution of two thousand dollars for the purchase of school books for the new school. The *Nome News* reported that she was a major benefactor in building a hospital, the town's first. That was the year that John Clum came to Nome as postal inspector and he and Earp renewed close friendships from their Tombstone days where Clum published the *Epitaph* newspaper.

Earp's reputation as a lawman was widely known and he was treated with respect. Although he was twice arrested for public drunkenness, neither of those cases was brought to trial.

The Earps left Nome for the winter of 1899, planning to open a saloon in Seattle. Wyatt and a partner opened the Union Club in Pioneer Square, operating it successfully as a gambling house until the State of Washington filed charges, confiscating the furnishings and burning them.

Returning to Nome in 1900, Earp sold his interest in the Dexter. When he and Sadie left Alaska, they had a bankroll of eighty thousand dollars—a small fortune. They first settled in the boomtown of Tonopah, Nevada, where they opened the Northern Saloon. They moved to California where they bought the first home they ever owned. Wyatt became an unpaid consultant for western movies and made many friendships with directors and actors. He died on January 13, 1929, at age eighty. His funeral service was well attended. Pallbearers were actors Tom Mix and William S. Hart; William J. Hunsaker, a well known Los Angeles attorney who had been Earp's lawyer in Tombstone; Jim Mitchell, a Hollywood screenwriter and reporter for the Los Angeles Examiner; George W. Parsons, a member of Tombstone's "Committee of Vigilance;" Wilson Mizner, a gambler and friend from the

Klondike Gold Rush; and John Clum. Mix was said to have cried openly during the service.

Sadie did not attend the funeral. Later she wrote that she was too devastated. Of Jewish heritage, she had Wyatt's body cremated and his remains were buried without public attention in the Marcus family plot at the Hills of Eternity in Colma, California. When she died in 1944, her remains were interred alongside his. The original tombstone marking their final resting place was stolen, but later it was replaced with a larger, more prominent marker. It continues to draw visitors who want to connect with the notorious gunman.

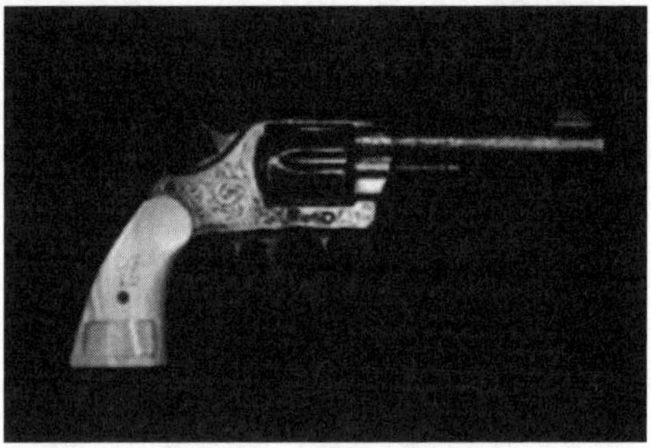

This revolver brought $35,000 to the city of Harrisburg, Pa., when sold at auction. Plaque on grip reads, "To Wyatt Earp, Welcome from the Citizens Committee of Nome." Another pistol reportedly left by Earp with the U.S. Marshal at Wrangell and unclaimed when the famous lawman's boat left early, is on display at the Red Dog Saloon in Juneau.

Many conveyances used in rush to get rich

Many conveyances were used by the hopefuls headed for the goldfields. Horses, dog teams, oxen, homemade boats, and even bicycles were used to propel men—and women—toward the creeks where they expected to find their fortunes.

Ed Jesson arrived in the Klondike in 1896, the same year George Carmack made the discovery that started the Klondike stampede. He ran a trading post at Star City, did some prospecting, and worked on others' claims but did not build much of a nest egg. When the rush faded after most of the promising ground in the upper Yukon had been staked and those with good color were being worked, he left Dawson on February 22, 1900, headed to Nome. The river was still frozen over. He did not have a dog team, but he did manage to acquire a bicycle.

Nine weeks later, he arrived at Nome after pedaling over icy, bumpy and occasional stretches overrun by water, for approximately 1350 miles. There is no known record of how many provisions he carried but it is presumed that he stopped at various roadhouses along the route.

Jesson boasted that he completed the trip safely, without experiencing a single punctured tire or broken spoke.

Another who powered his way on two wheels was twenty-year-old Max Hirschberg, who had first landed at Cook Inlet in Southcentral Alaska to join the miners at Hope and Sunrise. Hearing of the Klondike strike in 1896, he booked passage for Dyea and made it to Dawson. From there he elected to buy a bicycle and make his way to Nome.

Hirschberg's time for the distance is not known, but he did reach St. Michael and started across Norton Sound toward Cape Nome. His luck was not as fortunate as was Jesson's. Moving across the ice-clogged stretch, the chain which drove his wheel broke. Not about to turn back, the young man found a sturdy stick, jammed it inside the back of his coat and into his trousers, making a sail. Favoring winds propelled him toward the mouth of the Snake River, dodging clumps of ice along the way.

As with so many others who had "gold or bust" on their minds, the ingenuity exhibited by the cyclists surely stood them in good stead in their pursuit of wealth.

1900 Advertisement for Brighton Cycle and Motor Company
Photo courtesy Oldbike Museum

Frontier Nome struggles with population influx

Word of the discovery of gold on Anvil Creek on September 20, 1898, spread quickly after Jafet Lindeberg, John Brynteson, and Eric Lindblom contacted J.W. Price, Dr. A.R. Kittleson, and Johan Tornenson to form the Cape Nome Mining District. The "Three Lucky Swedes" had pulled into the mouth of the Snake River after bad weather caused them to turn back on a planned prospecting trip farther up the coast of the Seward Peninsula. It was a place where Lindblom had prospected earlier but found it to be only promising. Their pans on this late fall day, however, glistened with the yellow metal. They hurried back to Golovin to make things legally theirs by forming a district in accordance with the law.

Prospectors swarmed to the area as word spread, but found that claims had been filed with Kittleson, the district's recorder, covering the entire length of the creeks. Not only had the discoverers filed in their own names but had filed under powers of attorney for many friends and relatives. The extra claims were but the beginning of problems involving property rights that would plague the area for the next three seasons.

The discoverers named the river "Snake" because of its winding course and the rich main tributary after a hilltop rock formation that resembled the fixture upon which a blacksmith pounds red-hot iron. The name caught on and the camp that sprang up nearby was known as Anvil City. By the next summer, some three hundred prospectors had arrived, hoping to get in on the bonanza.

As it turned out to the benefit of many, other streams in the area produced prospects and a miner named Hummel discovered that even the beach sands would pay. There was plenty of reason to keep the hopes of the newcomers alive.

Across a stretch of Norton Sound to the south of Golovin was St. Michael, a seaport at the mouth of the Yukon River. It was there that steamers landed to disgorge adventurers who were headed to the fabled Klondike goldfields at the other end of the mighty river. It was there also from where two steamships in 1897 had departed and, upon reaching destinations in Seattle and San Francisco, sparked the great stampede.

It was Anvil City now that drew the world's attention, and there where the new horde would congregate. Within two years it would be Alaska's largest city, with a new name but facing all the challenges of a new frontier.

In the spring of 1899, ships anchored off the shore, a mile away from the beaches where tents were lined up in a steady line of white canvas. Passengers and freight had to be lightered ashore in smaller craft because of the shallow and rocky coast.

Buildings were raised quickly to house businesses geared up to serve people coming from around the world to get in on the excitement. Saloons, restaurants, grocery stores, hotels, and gambling houses—all with scales standing ready to weigh dust from the pokes carried by miners lucky enough to make a find, or working for those who had. Unfortunately, there were far more unlucky hopefuls than ones who found promising ground to stake or even to find work.

By October, after ice had closed the shipping lane and slammed the door on escape, crime was running rampant. Editor Fred A. Healy of the *Nome Daily Chronicle* lamented, "There is more felony in Nome than in the average city of half a million inhabitants." He blamed the lack of lights for the dangers faced by citizens. During the long winter nights, only the illumination from windows of saloons made it possible for people to find their way on the narrow board walkways lining the streets. Beyond those spots danger lurked in the dark.

An even worse problem was the lack of sanitation in a place crowded with humanity that was mostly male. In a published report on drainage and sanitation, Mayor Thomas D. Cashel announced that "The alleyways between some of the prominent saloons on Front Street are almost three feet deep with a glacier of urine."

According to the 1900 Census, Nome had a population of twelve thousand, four hundred eighty-six residents. There were one hundred and ten businesses including twenty saloons, four wholesale liquor stores, six bakeries, six restaurants, four hotels, twelve general merchandise stores, four bath houses, a brewery, a hospital, a bank, two printing firms, and a water works bringing water from the remote creeks to placer operations closer to the city. In addition to the traders there were sixteen lawyers, eleven physicians, two dentists, two surveyors, a mining engineer, and skilled workers in mining, construction, and transportation. There were two churches—one Protestant and one Catholic.

A year later there were one hundred twenty-seven lawyers—most of them busy handling matters dealing with mining claim disputes. Three more newspapers had been added and there was now a post office.

Editor Healy pushed hard for incorporation of the camp as a city, arguing that a formal organization was needed to provide essential services and protect the residents.

Meanwhile, the editor of the *Nome News*, John Franklin Alexander Strong, saw a need for a school. He argued that there were at the time twenty youngsters of school age and more were expected due to the steady influx of people.

An election was called to decide the question of building and operating a school, but there was little interest and those in opposition prevailed. That caused the Chamber of Commerce to offer to act in place of a government. A committee of business people attempted to come up with a plan, but competing interests and lack of funds prevented a satisfactory resolution.

"When the voters of Nome killed incorporation, they at the same time put it out of their power to select those who would govern them," Healy lectured on the pages of the *Chronicle*. "They now find that by their childish proceeding at the polls they did not do away with the need for government but simply transferred to the Chamber of Commerce their right of selecting the men who shall handle the public funds."

A new petition was presented to the court and another election was scheduled to be held on April 9, 1901, at the Golden Gate Hotel. Incorporation this time was approved by a margin of six hundred ninety-five to one hundred eighty-eight.

Elected to the city council were Tex Rickard, owner of the Northern Saloon; W. E. Geiger; J. B. Harris; S. H. Stevens, Jr.; W. H. McPhee; and Dexter Saloon co-owner Charles E. Hoxie. Also elected were school board members Miner Bruce, Colin Beaton, and Dr. J. J. Chambers. A property tax rate was set and an administrator, treasurer, and city marshal were hired.

A public library had been set up in the Congregational Church and that space was offered for a school. Miss Rosa Lamont volunteered her services as teacher without pay. Classes started under that arrangement, but before long a larger facility with more teachers was needed and a budget was adopted.

Nome children were occupied outside of school in helping their families meet expenses. One, five-year-old Franklin Munter, sold popcorn and sang songs for tips in the busier saloons. Harrison Loerpabel, age eight, had a route among the tents on the beach where he sold his mother's bread and pies. Moving later to international fame was six-year-old Jimmy Doolittle who sold the *Nugget* newspaper and delivered clean clothes from his mother's home laundry. Doolittle became the World War II general who led the first bombing raid over Japan.

The Nome school was operated by the city. Territorial schools were provided outside cities and supervised by Rev. Dr. Sheldon Jackson, general agent for education under the federal government. Jackson had established dozens of schools around Alaska, mostly operated in conjunction with religious missions despite recent legislation calling for separate facilities based on church-state relationships. Jackson's arrangement called for Native and non-Native children to attend the same schools, although that later was changed when the Bureau of Indian Affairs was given responsibility for educating Alaska Natives and assumed that function.

Nome's residents had plenty of opportunities for entertainment. The saloons were busy day and night, with live music for dancing. As in other mining camps, women were available as dance partners and were quick to share a glass of wine—for a price. The dance hall girls often had a sideline where extra attention was given—for an even bigger price.

Prostitution was illegal, but the law was not enforced. A discreet block off Front Street, screened by a fence, was known as "The Tenderloin" and was frequently visited by lonely miners who had a couple of dollars to spare.

The saloons often offered boxing matches featuring touring pugilists. Tex Rickart's Northern and the Dexter were among those where more exciting matches were held. Rickart went on to establish Madison Square Garden in New York City. Dexter co-owner Wyatt Earp was a referee who had been involved in the Fitzsimmons-Sharkey controversy not long before coming to Alaska.

The larger saloons had stages where plays were put on by visiting thespians as well as local talent. Burlesque troupes were also popular. Nome's first opera featured Broadway star Miss Emma Steiner who with Katherine Stagg wrote "The Little Hussar." Steiner, the contralto, was smitten with Nome and spent her summers from 1900 to 1910 mining on the Seward Peninsula. She was credited with discovering tin placers north of Nome.

Among the actors taking part regularly in Nome plays was author Rex Beach. Beach built a cabin at Rampart and tried his hand at mining before moving to Nome. He is known for writing "The Spoilers," a novel based on the real-life claim-jumping scandal that involved a crooked judge, a corrupt marshal, and a political king-maker from North Dakota who came up with a scheme to take over cross-filed claims and make off with the gold.

Civic organizations were plentiful. One, the Arctic Brotherhood, became a lasting gathering of Alaskans. Founded aboard the *S.S. City of Seattle* while en route to Skagway during the Gold Rush, it was a social order designed to provide top entertainment and celebrate participation in the grand adventure. “Camps” were established around the Yukon, including Nome.

Formal dances were held at various times, with Army officers from St. Michael and Nome’s Fort Davis among the elite who attended. Dance cards were filled out and attendees were dressed in their finery. A journal by the wife of Major Carroll Mercer of Fort Davis provides a good review of the activities. The major, whose family was politically connected, left the Army to take a diplomatic appointment.

A clipping from the *Nome Gold Digger* of July 10, 1901, quotes Mercer’s wife as describing in a letter to her sister a dance held by Nome elite:

“Saturday evening about fifty couples, Nome’s ‘400,’ danced a German (sic). It was the prettiest dance given here this winter, and was very exclusive. They had seven figures to open each dance—that is one couple opened a figure, and when all the people were on the floor that figure was completed and a new one begun. My partner and I opened one of the figures. The whole evening was a perfect success.”

Mrs. Mercer’s scrapbook from her days at Nome includes a dance card from April 15, 1901. The cover of the card notes that the dance was the “First Stampede, Nome Camp No. 9, Arctic Brotherhood.” It was held at the

Standard Theatre. The Order of Dance inside listed, "Grand March & Cicillian Circle, Waltz, Two-Step, Waltz Quadrillion, Rye Waltz, Schottiche, Minuet, Quadrille, Two-Step, Waltz, Newport, Lancers, Minuet, Two-Step, Waltz, Pollo, Quadrille, Comes, Rye Waltz, and Medley." Her card was filled out with a partner for each dance.

Situated on the exposed coast of the Bering Sea, Nome was battered by storms—as it continues to be today. Floods devastated the downtown and fires twice wiped out the pioneer city. The 1901 fire took out several businesses and residences with a loss of more than one hundred thousand dollars—multi-millions today.

The *Nome News* of May 29 reported on the fire under a double-deck headline:

ONE ENTIRE SQUARE
IS SWEPT BY FIRE

Fire Boys Work Like Heroes to
Check the Flames

A one-sentence lead paragraph of ninety-seven words—probably written by an adrenalin-fueled *News* editor John Franklin Alexander Strong—told both of the effect and the immediate rebuilding:

"Where only a few hours before had been a scene of wreckage and devastation where heaps of smouldering ruins and piles of twisted corrugated iron and rods had lain in heaps, in a few brief hours was a scene of great activity, new buildings going up on the ground, and where the sun had gone down on the scene of desolation and ruin, it soon rose to one of life and hustle, with the sound of the workman's hammer in the air and the shouts of the busy teamsters as they hurry lumber onto the charred grounds."

The story describes the fire-fighting effort, "With [only]

one stream of water the firemen were unable for some time to check the flames and they leaped from building to building with great rapidity." In eye-witness fashion, it goes through the struggle as "the firemen were beaten back, inch by inch," until "they took their final stand on the roof of the adjoining grocery store, and there stopped the ravages of the advancing fire king, with Chief Stanley at the nozzle."

Nome's first major fire on May 25, 1901, destroyed much of the downtown district. This photo was in the Eric A. Hegg collection. Note the ladder propped up to the window of a second-floor room as a rescue is undertaken. In the foreground, an onlooker shields his face from the heat with his hat. *University of Washington Libraries, Special Collections, HEGG 165B*

Author Rex Beach combines mining with acting

A free spirit who abandoned preparations for a lawyer's life to seek riches in the Northland became one of America's best-known authors. He returned to Winter Park, Florida, to become a major benefactor and alumni association president of Rollins College. He was given an honorary doctorate by the school—from which he did not graduate—and is buried in the front yard of the Alumni House.

Rex Ellingwood Beach was born September 1, 1877, in Atwood, Michigan. His family moved to Florida seven years later where his father, Walter, obtained land and established fruit orchards. Young Rex went to local schools, then enrolled at Rollins in 1891. He later enrolled in Chicago School of Law from 1896-97 and at Kent College of Law from 1899-1900.

A view of Beach's college days is given by Rollins College president Hamilton Holt when the honorary doctorate was awarded: Beach "was the type of student who does not permit studies to interfere with education." Records from Beach's college days indicate that "he was not always in synch with the Rollins administration of the time." In 1892, he was "severely reprimanded" by then Rollins president Edward Hatcher for "sailing on a Sunday" and the next week was suspended for "daring to attend a late night party in Orlando."

After cutting short his studies in his senior year at Rollins, Beach joined the Chicago law firm of his two older brothers. Admitting that he initially had dreams of being appointed to the United States Supreme Court, his view of a legal career faded. The clients traipsing in and out of his brothers' law offices, and the difficulty they had in coming up with payment for losing causes, gave him second, third, and fourth thoughts.

Hearing reports of great quantities of gold being found in the Klondike, the disillusioned law student made his way north in 1897, landing at St. Michael. With a grubstake provided by his siblings, he aimed to find claims around Dawson. The vessel he was taking was stopped by low water, however, and he spent the winter at Rampart where he built a cabin. He turned down a partnership with one man because of reports that the man had stolen a blanket. Finding another would-be miner who would go in on buying a claim on an off-shoot of Minook Creek, they set out to make their fortune. As it turned out, the first man hit it big and the chosen partner proved to be a scoundrel.

In his fictional short story "Pardners," Beach with tongue in cheek described Rampart thusly: "This here town was the same as any new camp; a mile long and eighteen inches wide, consisting of saloons, dance-halls, saloons, trading-posts, saloons, places to get licker, and saloons. Might not have been so many dancehalls and trading-posts as I've mentioned, and a few more saloons."

After two years chasing the elusive yellow metal, Beach had amassed a poke worth fifteen hundred dollars and headed Outside. Returning to continue his Yukon adventure, he and a friend elected to go overland by dog-sled and disembarked the steamer *Newport* at Katmai on the southern coast of Alaska. There they engaged an Aleut guide and set out for St. Michael—six hundred miles away as the crow flies, over a mountain range, facing streams and lakes and crossing unmarked territory. The grueling trip was almost fatal as Beach suffered a serious malady that left him weak and moving only with difficulty.

Reaching their destination in the spring of 1900, Beach was to witness the great influx that caused the Anvil Creek camp to incorporate as Nome and become Alaska's largest city.

He also was to live through the lawlessness of a frontier camp and view with a law student's eye the developing claim-jumping scandal. That scandal would be the basis of his start as a successful author. His novel "The Spoilers" was an immediate best-seller and would become a stage play and a movie version that would be produced four times. Another Alaska novel, "The Silver Horde" based on a Bristol Bay cannery battle, would appear first as a silent movie and again later as a talkie.

Beach had an uncanny business insight and kept the movie rights to his novels. The first release of "The Spoilers" movie gave him forty percent of the proceeds.

Gold Rush Nome also provided another prize. A troupe of performers came from New York to take part in plays and see what the excitement was all about. Most went back home, but one stayed behind to take advantage of the opportunities. Edith Greta Crater opened a hotel to serve the new arrivals. So many passengers in need of a place to sleep were being unloaded that she was forced to rent beds in eight-hour shifts and still could not accommodate all.

Beach, a playwright and actor, was smitten with the young woman. A natural blonde, she had dimples he said “looked like they were stuck into her cheeks by a mischievous finger.” They were to be married in New York City in 1907.

In his autobiography, “Personal Exposures,” Beach explained his nickname for her—The Wag Lady. A group of Nome hoodlums was known as the Wag Boys. The hotel owner had taken in an abandoned dog whose feet were injured and was in danger of freezing. She nursed it back to health, only to have two of the Wag Boys demand she give it back to them. The young woman told the girlfriend of the gang’s leader what had happened. How she had backed down the dog’s previous owner when he asked for its return and her love for the pet made an impression on the gang. The Wag Boys became her protectors for the remainder of the season.

Beach and his wife settled in Sebring, Florida, where he continued to write while pursuing an interest in horticulture. He was known for his varieties of gladiolus bulbs. They had a large house and owned seven thousand acres of land near Sebring and another five thousand acres near Indiantown.

Beach’s health had deteriorated and he suffered from throat cancer. Edith died in 1947 and her husband’s health worsened. He took his own life on December 7, 1949. Their ashes were mingled and buried outside the Alumni House at Rollins College. Beach’s manuscripts and other items are archived at the college where he is fondly remembered today.

Social activities keep Nome buzzing during winter

Nome suddenly became Alaska's largest city when thousands of would-be miners, adventurers, grifters, and curiosity-filled visitors flooded the muddy streets at the dawn of the Twentieth Century. Summertime offered work, travel, and sightseeing. Wintertime was a different story. In a place locked in by ice for half the year and where the sunlight was absent for long hours, loneliness was a problem. The many saloons were going day and night to draw in the thirsty and the gamblers. The larger watering holes also had stages where plays, concerts, and vaudeville acts were presented. Many exhibited boxing matches participated in by traveling pugilists who took on local toughs.

In addition to the businesses catering to an eager public there were fraternal organizations that held regular festivities. Before the settlement even became a city, there were more than a dozen social and philanthropic groups. Among them were the Elks, Ladies Aid Society, Arctic Chiefs, 75 Club, Pastime Club, St. Mary's Guild, Pioneer Dancing Club, Masonic Club, Nome Literary Society, Nome Cotillion Club, Eagles, Knights of Pythias, Loral Social Club, International Order of Odd Fellows, and others. Nome's newspapers were filled with articles—

including names of those attending—about dinner parties, dances, and theatrical appearances.

Music of what was to become known as the "Gay Nineties" came to the northland with musicians seeking to share the wealth. Songs such as "After the Ball," "Daisy Bell" (Bicycle Built for Two), and "The Band Played On" were featured. The "Can Can" and the "Hoochie Coochie" were favorites performed by female dancers who were hired to attract the miners and heighten their interest in spending a buck a dance, plus fork up for "champagne."

One of the most popular social groups in the northland was the Arctic Brotherhood. Organized on February 26, 1899, aboard the *City of Seattle* as it steamed from its namesake port to Skagway, a group of newspaper people, merchants, miners, gamblers, and adventurers wanted to celebrate their unique status as stampeders. Among them were Americans, Canadians, and people from other countries.

They adopted the slogan, "No Boundary Line Here." Their emblem was a gold pan with three nuggets and the initials A-B. A wide range of offices was established to oversee the organization. A uniform, with sash colors to illustrate the wearer's official title, and a set of bylaws were established. With tongues firmly in cheek, the passengers made up a formal initiation procedure under which a "chechaker" became a "bodyguard" to the "Arctic Queen," the patroness of the Brotherhood. The Creator was to be recognized with opening and closing prayers.

With a serious intended purpose, the hardy pioneers let their sense of humor shine through, usually by means of tongue-in-cheek expressions. Opening and closing songs were composed, to be sung by those gathered at their meetings.

The Camp Anthem was:

Hail, to our Arctic Queen;
Fairer there ne'er has been;
Hail to the Queen,
Hail to her Iciness!
Hail to her Mightiness!
She cometh now our camp to cheer.
Hail! Arctic Queen!

Allowing for expansion, the bylaws provided for "camps" to be established around the Yukon, with Camp Number One located in Skagway. From dues paid to the treasurer, whose post was identified as the "keeper of the nuggets," they purchased a lot and built a frame structure. Its false front was comprised of nearly nine thousand pieces of driftwood and flotsam gathered from the Skagway beach. That building, since refurbished, remains today on Skagway's main street and is called "the most photographed building in Alaska."

Nome's branch of the Arctic Brotherhood was Camp Nine. By 1910 there were more than thirty camps and the Brotherhood claimed ten thousand members, all tied to the Yukon.

Another benevolent organization active in Nome was the International Order of Odd Fellows. A national order, it was organized in 1819 in Baltimore, Maryland, to care for the sick, bury the dead, and aid widows and orphans. At Nome, the Oddfellows sponsored regular social activities in addition to its benevolent menu.

The Oddfellows chapter at Fairbanks was active for many years. Its two-story structure in downtown is now a retail store and is listed on the Register of Historic Sites.

Nome's elite included its wealthy residents and Army

and Revenue Service officers. Jafet Lindeberg, president of the Pioneer Mining Company and one of the Anvil Creek discoverers, built a mansion where he regularly entertained important visitors and hosted parties for local dignitaries.

Even though in a frontier town, Nome's social life was formal. Its dances featured formal attire, the ladies had dance cards, and orchestras provided the music. Those who attended were on their best behavior. Failure to abide by proprieties subjected offenders to banishment—a fate with consequences.

Leonhard Seppala with his team of Siberian huskies. Seppala won the last three Sweepstakes races, 1915 through 1917. He went on to become a national hero in 1925 when his team made the most hazardous and longest run in the relay of serum from Nenana to Nome to end an epidemic that broke out in the isolated city. Seppala was a founder of the Nome Kennel Club, with the intent of improving the breed of racing dogs. Photo courtesy Library of Congress

Dog sled racing brings lasting Alaska tradition

Far north outdoor activity was limited in the wintertime, but it invited spirited competition among those who owned dog teams. There were regular runs between the camps utilizing freight and mail routes; mushers vied for attention and boasted of the abilities of their dogs. The Pioneer Mining Company, owned by the Three Lucky Swedes whose discovery started the Nome boom, employed experienced mushers to conduct company business, hauling passengers and freight.

There was also much interest in improving the strain of dogs being used to increase stamina and prevent diseases. In 1907, the Nome Kennel Club was created to promote proper care of the animals and to seek ways to strengthen the sled dog breeds.

In 1908, enthusiasts established a race to show off the teams and drivers, the route to be from Nome to Candle and return, a distance of 408 miles. The route followed telegraph lines of the Army's Washington-Alaska Military Cable and Telegraph Service. It was patterned after the Kentucky Derby, with bugle, kennel colors, and all the hoopla connected with the celebrated horse race. Rules were rigid, with teams limited to ten to twenty dogs and mushers required to return with all dogs, including dead and/or injured. Dogs that pulled the sleds initially were of mixed breeds, ranging from native malamute sled dogs to setters and hounds. Siberian dogs were introduced soon after. Of course, betting was heavy with gold being laid on the table for the various teams. That first race was won by John Hegness, a native of Norway, who won the top prize of ten thousand dollars.

The following year, Russian trader William Goosak entered a team of Siberian huskies driven by Louis Thurstop, a Danish sailor. He did not win, but his team did so well it attracted much attention. Locals derided the smaller Siberians, referring to them as "Siberian rats," but admired their stamina and speed.

The following year Charles Fox Maule Ramsay imported seventy of the Siberian dogs and entered three teams, including the record-time winner. Ramsay, an English nobleman and the son of John William Maule Ramsay, the Thirteenth Earl of Dalhousie, was in Alaska to take part in the adventure and while in Nome became interested in dog racing. His three teams were entered in the names of himself and his two uncles. The team that won the race in the lasting record time of seventy-four hours, fourteen minutes and thirty-seven seconds was driven by John "Iron Man" Johnson, a Swedish Finn. Johnson also won the 1914 race.

The All-Alaska Sweepstakes was run annually until 1917, being discontinued as of 1918 due to the World War. The last three races were won by Leonhard Seppala, who was to go on to fame as the hero of the Serum run of 1925. Of interest is that the only American-born racer to win any of the original Sweepstakes was Fay Delzene of Iowa, who won the 1913 race in a time of seventy-five hours, forty-two minutes and twenty-seven seconds.

A seventy-fifth anniversary Sweepstakes race was run in 1983 and was won by Rick Swenson of Minnesota and a future four-time winner of the Iditarod Trail Race. Swenson's time was eighty-four hours, forty-two minutes and four seconds. The one hundredth anniversary race in 2008 was won by Alaskan Mitch Seavy. His time of sixty-one hours, twenty-nine minutes and forty-five seconds is the only one of the twelve Sweepstakes to best Johnson's 1910 time, being twelve hours, forty-four minutes and forty-two seconds faster along the same route.

'Scotty' Allan adds public service to racing

A nineteen-year-old lad from Scotland who in 1886 delivered a prized Clydesdale stallion to a rancher in South Dakota decided to remain in America. Experienced in caring for and handling animals, Allan Alexander Allan found work, married, and fathered two children.

He left his family in Oregon and joined the Klondike Stampede to seek his fortune. He prospected, trapped, and finally became a teamster, driving horses in the summer and dogs in the winter. Nome beckoned and there he entered the All-Alaska Sweepstakes, coming in first three of four years starting in 1909, placing second in 1910.

Allan became interested in politics and was elected to the Nome school board, then served two terms in the Alaska Territorial Legislature. In order to serve his constituents during the winter sessions in far-off Juneau, he drove his dog team from Nome to Fairbanks where he left them with a friend, then took a horse-drawn stage to Valdez, where he boarded a steamer for the capital city.

During World War One, the Allan-Darling Kennel sold to the French Army dogs for use in hauling supplies over the mountains as well as trained for sentry duty and search and rescue. The kennel owners were awarded the Croix de Guerre in recognition of their services.

Chugiak Artist Jon Van Zyle, known for his work with the Iditarod Trail Race, is shown with the 1909 Alaska Sweepstakes trophy won by Scotty Allan. The trophy ended up in California, among the possessions of John Marchant, a Nome gambler. Van Zyle had the broken handle reattached and presented the trophy to the Carrie M. McLain Memorial Museum. It joins five other of the eight race trophies on display there. PHOTO BY LEE JORDAN

Allan's 1909 trophy now on display in Nome

Scotty Allan's first-place trophy from the 1909 All-Alaska Sweepstakes was found in California where it rested for nearly a century. It has been returned to Nome and is on display at the Carrie M. McLain Memorial Museum with five others from the eight races. Leonhard Seppala's trophies from 1913 and 1914 are now on display at the University Of Alaska. The 1909 trophy was spotted at the home of James and Betty Harrah where it was ignominiously used as a door-stop and umbrella stand. Marjorie Cochrane, a former Alaska newspaper reporter, saw the trophy and inquired about its history. The author of a children's book. "Three Dogs, Two Mules, and a Reindeer," she contacted its illustrator, Chugiak artist and former Iditarod racer Jon Van Zyle, who offered to repair its broken handle and present it to the museum.

As related by James Harrah, the trophy was found among items stored in a building purchased from the estate of John Marchant. Marchant was a gambler at Nome during the Gold Rush before returning to San Francisco to publish a horse racing newspaper. The trophy apparently was put up to cover a bet on a losing poker hand. Not aware of its significance, Harrah had no particular interest in the 18-inch-tall, five-pound solid silver cup.

Fay Dalzene was the only winner of the All-Alaska Sweepstakes who was born in the United States, hailing from Iowa. His time in the 1913 race of 75 hours, 42 minutes and 27 seconds was the second-fastest of the original series.

Glenbow Museum photo 238156

Iditarod Trail Sled Dog Race honors Serum Run

A Gold Rush figure in Nome, the most skilled of the many dog mushers there, each year holds the honorary Number One position among the scores of people who take part in the annual Iditarod Trail Sled Dog Race. In the ceremonial start, a dignitary is chosen to begin the race to remember the heroic musher's part in the 1925 serum run to aid diphtheria victims in Nome.

Norwegian Leonhard Seppala knew Jafet Lindeberg, one of the discoverers of gold that started the Nome Gold Rush. On a visit to his home country, Lindeberg convinced Seppala to go back and work for him in the young boom town. As an employee of Pioneer Mining Company, Seppala cared for their kennel of sled dogs used to haul supplies and mail.

When Roald Amundson, a Norwegian explorer, attempted to cross from the Atlantic to the Pacific via the Arctic waterways in 1906, he was stopped by ice and had to turn back. In anticipation of Amundson's planned stop in Nome, Lindeberg had obtained puppies from Siberia as a gift for the noted personage. After the visit was thwarted, the miner gave the pups to Seppala, who raised them and trained them to pull sleds.

Entering the 1914 All-Alaska Sweepstakes, Seppala

and his young team traveled along a trail neither he nor the dogs had used before. They almost went over a ridge and into open water, barely avoiding disaster. Seppala withdrew from the race. He was to come back the next year, winning it and the next two of the final contests to be held in the original competitions.

In 1925, when a diphtheria epidemic struck Nome in the midst of winter, the hospital found its supply of serum not only very limited but outdated. The nearest dependable quantity of serum was at the Alaska Railroad Hospital in Anchorage. Plans were made to rush serum by train to Fairbanks, from there to be transported by dogsled in relays.

That choice was controversial, with aviation enthusiasts advocating for the fledgling bush airplane service to be used. It was decided that dogs were more dependable as fuel supplies were uncertain and the airplane had not yet been proven in the harsh winter conditions expected to be encountered.

With quick delivery deemed critical, dog teams going overland around Norton Sound would reuire an extra day of travel. Organizers of the mercy run turned to Seppala, who had crossed the Sound on more than one occasion and was the driver most likely to cross safely—and the most competent to make the decision to cross or avoid the ice-covered bay. He traversed the Sound safely and proceeded at top speed to Nulato, four hundred miles from Nome. He almost missed the relay when the musher with the serum was ahead of schedule and passed without being recognized. The incoming musher hailed him and made the transfer.

Even though his dogs were tired, the wind chill factor was minus eighty-five degrees and gale force winds were howling, Seppala turned his team around and headed back to Nome, stopping only briefly along the way. He drove the team another two hundred and sixty miles to Golovin where he met Charley Olsen. Olsen, whose hands had been frostbitten in the high winds, passed the precious vials, each with three hundred units of serum, to Gunnar Kaasan who completed the dash to Nome. The serum was soon thawed and administered, diminishing the epidemic's threat.

The national press lauded Kaasan and his lead dog, Balto, who were later welcomed as heroes by an admiring population when they toured the States. Seppala, whose far superior accomplishment was overlooked, felt his lead dog, Togo, deserved commendation for his outstanding performance. The eleven-year-old dog had led the team for one hundred and seventy miles in three days before turning around to begin the final dash. As they crossed the Sound at night and in a blizzard, Seppala said he could not see the trail, but Togo safely led the team to a roadhouse at Pilot's Point. From there they climbed a five-thousand-foot mountain before descending to Golovin where they handed off the serum to Kaasan. Togo's racing career was ended by the rigors of the mercy run. He did gain recognition from the explorer, Amundsen, who presented the heroic dog with a medal in a ceremony at Madison Square Garden.

First used at the turn of the Twentieth Century, the Iditarod Trail linked Nome with the open-water port of Seward in southcentral Alaska. The annual race, which was inaugurated in 1972, roughly follows that trail and is dedicated to the mushers who delivered the serum.

Cutter *Bear*, Capt. Mike Healy play big role in Arctic

While not directly involved in searching for the yellow metal that brought thousands to the north country, one particular ship and her master were very important to the people who were.

Plying the waters of the Bering Sea long before the Klondike became the focal point of the world was the United States Revenue Cutter *Bear.* Designed to be a part of the sealing fleet, the ship launched in Scotland in 1874 spent her first decade in that trade, operating off Newfoundland. But when the Arctic expedition of Lieutenant Adolphus Washington Greely became stranded, the Navy prepared a rescue effort. The *Bear* and two other ships, *Thetis* and *Alert*, were chosen because of their sturdy construction designed to withstand ice. The vessels were loaded and dispatched to the Arctic.

On June 22, 1884, the *Bear* sighted the surviving members of the expedition. Without adequate provisions following a series of misfortunes, the men were left without shelter or transport when winter fell upon them. They subsisted on whatever game might be found, moss, and even leather gear. Several died, others were driven mad and even their surgeon committed suicide. The survivors when found were described as mere skeletons.

Despite the heroic outcome of the rescue, the *Bear* was decommissioned by the Navy and turned over to the Treasury Department for use in waters off Alaska. That began a forty-one-year career as a Revenue Cutter—a record that has never been matched.The first captain for the *Bear*'s maiden voyage as a revenue cutter was A. A. Fengar. Within the year she was put under the command of the man whose name would almost become synonymous with that of his ship.

Michael Augustin Healy was born in Macon, Georgia, in 1839. His parents were Michael Morris Healy, a plantation owner, and Mary Elisa Smith, a former slave. The couple had ten children, the one to gain renown for his seamanship the fifth among them. Because they were of mixed African and Caucasian blood, the children were sent north to be educated in Catholic schools. Three brothers became priests; one, James, elevated as the first black bishop in North America. Patrick was to become the president of Georgetown University and Sherwood was an expert in canon law. Three sisters were nuns, one of them a mother superior.

The son who was to become known as "Hell-Roaring Mike" shunned both academia and the cloth, turning to the sea. He shipped out at age 15 as a cabin boy on an American-flagged East Indian clipper. In succession, he became an expert seaman and was commissioned as a mate on merchant vessels.

In 1864, Healy applied for a post in the Revenue Marine and was accepted as a third lieutenant. He served on several cutters before being assigned as second officer on the cutter *Rush,* destined for Alaska waters. His first command was on the *Chandler* in 1877. His next command was on the *Thomas Corwin* in 1884.

It was two years later that Healy became the skipper of the *Bear* for its first entry into Alaskan waters. It was a post he was to hold for nine years, encompassing some of the most important events in history. He would transport John Muir on scientific missions, take government officials to their posts, and patrol the waters to control smuggling and look after the nation's interests. The *Bear* became involved in the program to introduce reindeer into Alaska and transported Dr. Sheldon Jackson, the federal superintendent of education, to various coastal communities.

When taking federal marshals and judges to different locations, the *Bear*'s deck often became a courtroom. In many instances, there was no civil authority on board and Healy became the one who took care of legal problems. He was known as a man with a quick temper, but fair. On southbound trips, the *Bear*'s brig held prisoners destined for federal prisons or the mentally ill to be treated in Oregon's Morningside mental institution.

The *Bear* was called upon to make another major humanitarian trip in March of 1898 when whaling ships were trapped in the ice and were running out of supplies. The cutter was unable to reach the stricken vessels, but two of her officers were designated to lead the Overland Relief Expedition which drove herds of reindeer to feed the stricken whalers. (See the Reindeer Section following this portion of the book.)

During World War I, because of its construction designed to break through heavy ice, the *Bear* was taken over by the Navy and used in Alaskan waters. Decommissioned, she again was brought into the nation's spotlight when Admiral Richard E. Byrd arranged to obtain the moored vessel which was being used as a maritime museum in California. It was fitted out and used to take two different expeditions led by Byrd to the Antarctic.

Still seaworthy, the *Bear* was again taken over by the Navy and pressed into service patrolling the waters off Alaska during World War II. After more than two-thirds of a century of weathering icy winds and pushing her way through ice floes, the gallant *Bear* was finally sidelined for good. She was acquired by a Canadian firm with the intention of again turning her into a whaling ship but she remained tied to the dock when the price of baleen and oil no longer made it feasible to harvest the giant mammals.

One more attempt to immortalize the valiant vessel was made when Alfred M. Johnston of Villanova, Pennsylvania, purchased her. He planned to moor the ship at Philadelphia and open her to the public as a restaurant and museum. It was not to be. While under tow, her forward mast collapsed, the ancient timbers in her hull gave way, and *Bear* settled to a watery grave in the depths of the Atlantic.

Part III
Reindeer Introduced for Alaska Natives

Sami reindeer herders milk a reindeer at Port Clarence in this 1900 photo. Rev. Sheldon Jackson acted on a suggestion by Capt. Mike Healy and introduced the animals as a way to supplement the diet of the Native peoples and also provide an industry. Reindeer are still raised on the Seward Peninsula by Native herders. Photo by Eric A. Hegg

University of Washington Libraries, Special Collections, HEG 1347

Sheldon Jackson brought schools, reindeer industry

A missionary who worked to establish schools in many undeveloped areas of the United States joined forces with Captain Michael Healy to create an industry for Alaska's Natives.

Sheldon Jackson was born in 1834 in Minaville, New York. His mother, Delia Sheldon Jackson, was the daughter of Alexander Sheldon, Speaker of the New York Assembly. Her son attended schools in his hometown, then graduated from Union College in 1855. He completed Princeton Theological Seminary in three years. He was ordained by the Presbytery of Albany as a Presbyterian minister on May 5, 1858, and promptly married Mary Vorhees.

The couple embarked on a career as missionaries with the young minister's first assignment to the Chocktaw Mission in Oklahoma. They were forced to return to the East because of the young man's ill health. Restored and ready to return to work, Jackson's next appointment was to LaCrescent, Minnesota. In the next decade he was credited with organizing twenty-three churches.

Jackson took advantage of the advancement of railroads across the country, his travels taking him by train and stage to Idaho, Colorado, Wyoming, Montana, Utah,

Arizona, Mexico, Texas, and Nevada. He was said to have established a church at every stop—in all more than one hundred.

In 1877, the forty-three-year-old minister ventured to Alaska where he planned to establish schools and traveled extensively, preaching and organizing schools in the southeastern Panhandle. There he became dedicated to bringing education to the entire territory that had been acquired from Russia just ten years earlier.

Traveling to the nation's capital in Washington, he visited President Benjamin Harrison, a fellow Presbyterian he knew from both church and political connections. Having seen conditions in the new frontier, he was interested in arranging for local government and worked with officials to help draft legislation that would bring it about. He was instrumental in getting the Organic Act of 1884 passed by Congress and signed by the President.

The following year, Jackson was appointed as Special General Agent of Education for Alaska. He had a budget of $25,000 and authority to establish schools and set policies.

Based on his experience with schools on the Indian reservations, Jackson formulated directives to be issued to teachers. He followed the practice of "deliberate acculturation" whereby students would be taught only in English with use of their native tongue not only discouraged but prohibited. His religious beliefs were emphasized by insisting that spirituality be included in the teaching.

Recognizing the spiritual aspect of education also had another purpose. Alaska was a huge and diverse area and centralized control would be both difficult and very expensive. Jackson knew that missions had been set up in many communities by various religious denominations. If those missions could operate schools under his jurisdiction, it would benefit the government as well as the missions whose teachers would draw salaries and supplies paid in part from Jackson's budget.

He chose to apportion sections of Alaska to the various denominations, based primarily on existing conditions. Cook Inlet and Kodiak Island would be assigned to the Baptists. The Yukon River would be served by the Canadian Anglican Church and Episcopalians. The Aleutian Islands would be served by the Methodists. The Kuskokwim River and its Delta would go to the Moravians. Prince of Wales Island and surrounding area would be served by the Congregationalists. Kotzebue was assigned to the Friends Church. The Seward Peninsula would be covered by the Lutherans and Covenant Church adherents. Southeast Alaska, where the Presbyterians were already established, would be theirs as would St. Lawrence Island and the northern coast.

Noteworthy is that the Russian Orthodox and the Roman Catholic churches were to be "encouraged to continue their work in educating children" without being assigned geographical areas. Some of Jackson's writings indicate that he had a prejudice against these denominations within the Christian churches.

Jackson's policies were put into effect, but with unforeseen consequences. Native families objected to being told to abandon their customary potlaches, dances and traditional celebrations. It was an affront to see their children punished for speaking their native tongue; elders who feared their culture would be lost were distressed. That fear was realized decades later when it was found that the number of people who were conversant in the various dialects was dwindling to the point of extinction.

While the General Agent of Education supported acculturation*, he was opposed to taking students away from their homes and families and sending them to distant schools.

There also was criticism from some who objected to public schools being operated by missions and paying teachers from public funds. That argument later led to an amendment to the Indian Appropriation Act of 1893 that segregated schools, specifying that Alaska Native pupils were to be taught in Bureau of Indian Affairs schools.

While Jackson's official attention was directed toward education, his missionary instincts made him worry about the welfare of the Native population. Their subsistence lifestyle was subjected to many things beyond their control. He sought ways to substitute a more dependable source of sustenance.

*In 2012, the Presbytery of Yukon presented an official apology to the elders of Saint Lawrence Island for policies that discouraged use of the Native language and customs.

Reindeer project adds jobs, aids stranded whalers

During his voyages with Captain Michael Healy on the Revenue Cutter *Bear*, Rev. Sheldon Jackson learned that on the Siberian side of the Bering Sea the natives raised herds of reindeer that were used for transportation and for food. Jackson considered this as a possible industry for the Alaska Natives. With support from Healy, Cape Prince of Wales missionary William Thomas Lopp—who was familiar with reindeer herders in his native Norway—and from the manager of the Alaska Commercial Company, Jackson petitioned Congress and gained permission to import reindeer to establish herds—but no money for the purpose. He turned to his church and received two thousand dollars in individual donations. As an experiment to test viability of the idea, he bought sixteen reindeer for a pilot project and found that they not only survived through the winter but produced two calves.

Healy made five more voyages to Siberia and brought one hundred seventy-one deer and five herders to the Teller Reindeer Station at Port Clarence. The Siberian herders were supposed to teach the Alaskans to care for the animals, but the men were not able to get along. Healy and Jackson had not taken into account that the people who lived on opposite sides of the Bering Strait traditionally had fought. As it was, the Alaskans traded with the Siberians for reindeer skins and other items. It was an advantage the Siberians did not want to relinquish.

In 1894, Jackson appealed to Scandinavians by advertising in American newspapers catering to those readers. There were two hundred and fifty responses to ads seeking applicants to come to Alaska to train people to herd and care for reindeer. William Kjellman was one who responded and was selected to be put in charge of the Teller station. A Lapplander who spoke the Sami language, he was sent to Norway to choose men who could teach the Natives to care for the animals. He contracted with herders who would come to Alaska with their families, dogs and sleds to take on the task.

Alaskan apprentices were signed up to enter a five-year program of training. Each year they were given two female deer to care for. At the end of the apprenticeship, they were loaned enough additional animals to bring their herds up to a total of 50.

In 1906 a government study reported that the program was being abused and that the majority of the animals were owned not by Natives but by the Scandinavian instructors and missions where herds were headquartered. One of the chief critics was William Lopp, Jackson's protégé. Jackson was replaced by Lopp and government employees took over the reindeer management program.

Teacher William T. Lopp replaces Jackson

William Thomas Lopp was born June 21, 1864, in Valley City, Indiana, where he attended local schools before in 1888 earning a degree from Hanover College. He taught school and served as principal until seeing an advertisement seeking mission teachers in Alaska. In 1890 he was assigned to a post at Cape Prince of Wales. Because of reports that the Natives there were troublesome, a second teacher, Harrison Thornton, was sent to accompany him.

The two teachers arrived at the mission that summer and set about building a school with help from a ship's carpenter. Cape Prince of Wales is located on the westernmost point of the Seward Peninsula on the shores of the Bering Strait. It is frozen in and isolated for about seven months of the year. The only means of travel in that time was by dog sled or walking with snowshoes.

As it turned out, the Virginian Thornton's personality was unlike that of Lopp, who was six years younger. Lopp was interested in learning the Native culture and language while Thornton was stand-offish and rigid in his approach. Although he had boasted of being an outdoorsman, Thornton proved to be of little help when taking part in a hunting trip with villagers. He in fact had to be rescued in one instance, a situation that put his companions in danger. As their time at the Cape lengthened, Thornton was to become more withdrawn and fearful.

In 1892, a third teacher was assigned to the mission. Ellen Louise Kittredge was born March 10, 1868, at Westboro, Massachusetts. She was living in Minnesota when she enrolled in college at the age of fifteen, supporting herself partly by teaching, and in 1889, she was principal and head teacher of the school in Glyndon. As had Lopp, she saw an American Missionary Association appeal for a teacher to travel to Alaska, this time in company with a husband and wife who were bound for the mission at Cape Prince of Wales. That couple was Thornton and the woman he married while on vacation that summer.

Before the Thorntons' first wedding anniversary, he would be murdered and his wife a widow. Three teenagers, angered by Thornton's affronts and bolstered by home-made liquor, knocked on his door one night. When they heard the man unlatch the door, one fired a whale gun and a second let loose a rifle round through the door. The missionary was thrown back, a large gaping wound in his mid-section, as his horrified bride looked on. The three young men were captured by village elders and themselves killed as punishment.

On their first meeting, Ellen was impressed by Tom Lopp. They soon fell in love and were married. They were to have eight children who were brought up with the village children. The eldest was Lucy, a sickly child who died before reaching majority. Their second child, a son named Dwight, was brought into the world assisted by a Native woman, Mary Antisarlook, the wife of a reindeer herder. Charlie Antisarlook was to soon play a major role, along with Lopp, in delivering reindeer to Barrow to aid the crews of eight whaling vessels that had been caught in ice and stranded.

Lopp was convinced that Sheldon Jackson's proposal to import reindeer to supplement the caribou as a source of meat was of merit. He soon left his position as teacher and was named superintendent of the Teller reindeer station, replacing Miner Bruce, who had been removed due to complaints of improprieties. He later was put in charge of a new reindeer herd at Cape Prince of Wales.

It was in this latest capacity that he was asked to take part in the Overland Relief Expedition.

When it became known that the stricken whalers might be in danger of starvation, the newspapers in San Francisco pressed hard for the government to bring relief. Much of the whaling fleet was based in San Francisco, which also was the home of the owners of the Yukon trading posts. Two leading newspapers picked up the cause, each competing to be the champion of a humanitarian action. Wanting to curry favor in the West, President William McKinley told his Secretary of the Treasury, Lyman Cage, to have the Revenue Marine Service mount a relief expedition. That Cage conferred with Jackson is apparent from the detailed instructions given to Captain Francis Tuttle, commander of the Revenue Cutter *Bear*. The *Bear* had just returned from its Alaska duty and was ordered to immediately be refitted and return north.

Tuttle headed back to the Bering Sea and went as far north as he could before encountering ice. He pulled in at Cape Vancouver and put three members of his crew ashore with orders to acquire dogs and go overland to the Seward Peninsula where they were to ask Antisarlook and Lopp to provide reindeer and drive them to Barrow. First Lieutenant David Jarvis, Second Lieutenant E. P. "Bully" Bertholf, and Surgeon S. J. Call were landed on December 15, 1897. Bertholf was to go to Cape Nome to contact Antisarlook while Jarvis and Call were to go to Cape Prince of Wales to contact Lopp.

Both herders agreed to surrender their deer and go with them to Barrow—under strict conditions that the reindeer be replaced.

The arduous trek of fifteen hundred miles was completed at the end of March, 1898. It was learned that most of the whalers had survived, although several had come down with scurvy. Starvation had been avoided because the supplies on the stranded ships had been combined with supplies available at Barrow and rationing was effective. The herds were of benefit, however, because the fresh meat helped fight off scurvy and revitalized the survivors. In celebration, the rescuers, whalers, and villagers enjoyed a baseball game on the ice in May. The *Bear* broke through to Barrow on July 28 and Tuttle welcomed his officers back on board.

Lopp returned to Cape Prince of Wales where he was reunited with his family. Lopp argued that more of the animals should be in the hands of the Natives, saying too many had been given to the Lapp herders and missions. In 1904 he was appointed as superintendent of government schools and also replaced Jackson as head of the reindeer project.

Leaving Alaska and taking his family to Seattle, Lopp continued to serve both actively and as a consultant. For a period he was hired by the Hudson's Bay Company as a reindeer consultant. He died in Seattle on April 10, 1939. Ellen Lopp lived for several years afterward. Her granddaughter published a book titled "Ice Window."

Part IV

Felix Pedro find opens Fairbanks district

Miners show off their clean-up at a creek in the Fairbanks Mining District. Photographer Charles F. Cann

P341-Cann-5 Alaska State Library Photographers in Alaska Photo Collection

Barnette's fortunate landing turns into misfortune

The founding in 1902 of what was to surpass Nome as Alaska's largest city, and remain today its third-largest, came after a series of misadventures. One Elbridge Truman—or maybe Ezra Thompson—Barnett was in Helena, Montana, in 1897 when he heard of the Klondike discovery. Preferring to be known only by his initials ET, his actual birth name is not known.

How Barnette came to be in Helena, or where he lived since his birth in 1863 in Akron, Ohio, with one exception is not known. The exception is that in Oregon at age seventeen he was brought up on charges. A partner in a horse-trading venture that was to take place in Canada convinced the court that Barnette stole money from him. Barnette served one year of a seven-year sentence before being released on the condition that the young man never again show his face in Oregon.

Arriving at the mouth of the Yukon, Barnette formed a partnership with other stampeders to buy the steamer *St. Michael* and take it to Dawson. Barnette was nominated to skipper the vessel and thenceforth was known as "Captain" Barnette. The skipper was a smooth talker, but whether he was an industrious visionary or a villainous schemer will have to be left to the reader's opinion.

The *St. Michael* got only as far as Circle City before being iced in. It had left late in the season and encountered low water and sand bars, causing delay that prevented it from reaching its destination prior to freeze-up. Barnette bade farewell to his passenger-partners, leaving them to fend for themselves. He arranged to get a sled and dog team, speeding upriver in hopes of being among the lucky ones who would file on rich claims. Alas, by the time he arrived, all the known promising creeks had been staked.

Convincing officials of the North American Trading and Transportation Company that he could manage their mining properties, Barnette was paid enough for his winter's effort to cover his way back to Helena. There he married Miss Isabelle Cleary and convinced her to go to Alaska with him. Her brother, Frank, accompanied the couple.

In 1901, Barnette formed another partnership, this one with Charles Smith, who he had met at Circle. During his time at Dawson, Barnette had come to know trader John Healy, who envisioned an All-American route to the interior of Alaska. Where Healy's proposed route would intersect the Tanana River would make a good location for a trading post, Barnette reasoned.

Barnette and Smith arranged for twenty thousand dollars' worth of supplies to be shipped to St. Michael. The steamer *Arctic Boy* was purchased to haul them and their goods upriver to the location of their proposed trading post. Unfortunately, Captain Barnette's navigational skills were not on par with the challenges of the Yukon and the steamer had to be beached. Stymied, the captain arranged to sell the supplies. Before the cargo was transferred, however, James H. Causten, a customs inspector, invested six thousand dollars on the promise that he would get one-third of the profits from the proposition. Barnette used that investment to reclaim the supplies he had sold.

Another steamer, the *Lavelle Young,* was obtained to take the goods to the intended location. It, too, ran into low water and sand bars that necessitated ferrying the cargo in stages. Finally, after making it to the Chena River tributary but unable to go farther, Barnette decided to build a shelter for the store and a cabin for his wife and himself. He planned to spend the winter there, then make arrangements to transfer the goods farther up to his original intended destination.

But his first customer turned out to be someone who changed Barnette's life and the course of Alaska history.

That customer was an Italian named Felice Pedroni, who with his partner Tom Gilmore were headed for the Tanana Hills in search of a creek where gold had been found some years earlier. Pedroni went by the name Felix Pedro, hoping it would be easier for the Americans to pronounce.

Renaming the spot where Barnette's Trading Post sat as Chenoa City, the Barnettes went to Seattle that winter to obtain more supplies and have a smaller boat built. Naming the boat *Isabelle* for his wife, the captain planned to equip it with the steam engine and other fixtures from the beached *Lavelle Young.*

Returning by way of St. Michael, Barnette found that the new boat could not handle the Yukon's obstacles any better than its predecessors. He was able to restock the trading post's inventory and continue operating there.

During his brother-in-law's absence, Cleary grubstaked Pedro. Shortly before the Barnettes arrived, the prospector came by and announced that he had found a rich prospect. Cleary, Barnette, and Pedro joined to form a district where their claims would be formalized.

While at St. Michael Barnette had visited with Judge James Wickersham, the judge asking that an Indiana friend, Senator Charles Fairbanks, be honored by having his name attached to something. That request was honored when the new district was recorded as the Fairbanks Mining District. Barnette was named as recorder.

With an eye toward promoting his trading post, Barnette made sure that newspapers were aware of the new strike. During the winter of 1902-03, hundreds of stampeders arrived. Ever on the lookout for financial gain, the trader hiked up the price of flour to three times its former amount, then limited buyers to one sack—plus requiring that the newcomer also buy three cases of canned goods. Faced with threats to lynch the trading post's manager, Barnette backed off on the bundling proviso, but kept a higher price on the flour.

In 1906, James Clausen sued Barnette, claiming that the trader had reneged on his promise to pay him the promised one-third share of profits. He wanted a third of everything Barnette had made since the six thousand-dollar advance five years earlier. The defendant argued that Clausen was due only a one-third share of limited profits. Wickersham ruled that Barnette owed only a third of profits made from his Tanana Valley dealings.

During the trial, the Oregon sentence was made public. Combined with ill feelings over his early price gouging, public opinion was turned against the man who was elected as Fairbanks' first mayor. In that election, Barnette's opponent garnered sixty-seven more votes, but the city council was arm-twisted and declared Barnette the winner. That fact did nothing to improve public opinion regarding Barnette's character.

In 1909 Barnette purchased the Washington-Alaska Bank and the next year bought the Fairbanks Banking Company, making it part of the other institution. Four days into the new year 1911, the Washington-Alaska Bank filed bankruptcy and suddenly closed its doors. Depositors were said to be left holding the bag in the amount of a million dollars. In the middle of the night on March 22, the banker skipped town, believed to be carrying half a million dollars in bank funds. Arrested in Los Angeles, Barnette was charged with embezzling fifty thousand dollars, only ten percent of the money he was thought to have with him when he absconded.

In a trial held in Valdez, Barnette was acquitted of the embezzlement charge, but found guilty of "falsifying a financial document." He was fined one thousand dollars. The *Fairbanks Daily News-Miner* complained in an editorial, "(The decision) was the rottenest judicial farce the North has ever faced."

The Fairbanks climate no longer suiting him, Barnette moved Isabelle and their two daughters to Los Angeles. Isabelle not long afterward divorced her fifty-five-year-old husband after finding love letters from another woman. She was given custody of the two girls. Barnette went to Mexico where he lived on a palatial estate. He died on May 22, 1933, in Los Angeles after being injured in a fall. Isabelle died in September, 1942, in San Jose, California.

His Alaska legacy tarnished, even the Barnette's Cache name given to his trading post now exists only in the history books.

Felix Pedro discovery makes Tanana next boom

Born on April 16, 1858, as the youngest of six brothers to a poor family in Fanano, a town in the Duchy of Modona, Italy, Felice Pedroni left home at the age of twenty-three following the death of his father. Forty-four years later, he was to be the person responsible for Fairbanks for a time becoming Alaska's largest city.

Immigrating to the United States in 1881, Pedroni adopted the name Felix Pedro. Little is known of the man's life other than that he worked his way across the country. For a dozen years, he went from one town to another, pausing to make enough money to get him to the next stopping place.

How he got to Alaska, as well as exactly when, remain unclear. It is known, however, that he was in the Circle area prior to the rush to Birch Creek in 1893. He was based there as he prospected creeks that flowed into the mighty Yukon River. He was looking for gold in the Fortymile district in 1898 when his attention was concentrated on the Tanana Hills area. At the time, thousands were stampeding to the Klondike and the Three Lucky Swedes were to make their find on Anvil Creek, at the other end of the Yukon.

It was that year—destined to become famous worldwide—that the Italian and a partner found a good prospect. Winter was approaching and their food was running dangerously low. Pedro planned to return the next spring and tried to memorize landmarks. Whether it was worry about survival or some other concern, those landmarks were not sufficiently imprinted on his brain. He and Tom Gilmore, a miner also based in Circle, searched for the next two summers but without success. Pedro was undeterred, feeling that great wealth was at hand.

In the spring of 1902, the two were back in the area. Climbing to the top of Ester Dome, they spotted a plume of smoke coming from a steamboat making its way upriver. They made out the name *Lavelle Young* as the trader's boat pulled into an eddy below. Hurrying there, they replenished supplies. It was around June 22.

The two left Barnette's store and headed north-northeast, arriving at the junction of Goldstream Creek and another stream which they named Gilmore. Their pans showed colors and they began digging a shaft. It was warm and it was some distance before they encountered permafrost. Surprisingly, they hit bedrock at only fourteen feet. Quickly scraping bedrock and shoveling the gravel into a pan, they washed away the muck and found the bottom of the pan covered with gold.

Pedro and Gilmore filed claims for themselves and proudly informed others of their find. Barnette was thrilled at the prospect of a stampede that would bring customers. Pedro raised a tent and invited Gilmore, Barnette, and others in to organize a district. Barnette remembered his conversation with Wickersham and named the district Fairbanks in honor of the judge's friend, the Indiana senator.

Pedro continued to make his home at Circle where he

had fallen in with an Irish woman named Mary Doran who lived with him as his wife.

Pedro suddenly fell ill and in 1910, at the age of fifty-two, was rushed to St. Joseph's Hospital in Fairbanks. He died there on July 22. It was thought he had succumbed to a heart attack, but a friend from the old country, Vincenzo Gambiani, was suspicious. He claimed that Pedroni had never exhibited symptoms of heart trouble. Pedro's embalmed body had been sent to San Francisco and was buried in Coloma.

On his deathbed, Gambiani again insisted that his friend had been murdered. Unable to speak, he wrote, "Poison. Wife." In October of 1972, Pedroni's remains were exhumed and sent to Italy. An autopsy was performed and the results of hair sample tests were said to indicate that he was indeed murdered. His remains were reburied at Fanano. It is not known if Mary Doran or anyone else was charged with his death.

A monument was erected at Mile 16.1 of the Steese Highway, near Pedro Creek, the site of Pedro's discovery. The profits of his mining had long been dissipated and he died a pauper. His legacy, however, has lived on. A stone monument was presented by the people of Pedroni's hometown which became a sister city if Fairbanks. In recognition of Pedro's contribution, the monument is rededicated each year in opening ceremonies for the annual Golden Days celebration in Fairbanks.

'Cassiar Angel' Nellie Cashman gains Alaska fame

An Irish woman who was "as pretty as a picture but tough as nails" made a name for herself in the Old West, the Klondike, and Alaska. She was a business woman, a miner, a musher, a benefactor, and an adventurer.

Ellen "Nellie" Cashman was born in Queenstown, County Cork, Ireland, in 1845. Her family name was O'Kissane but anglicized to Cashman after her mother brought Ellen and her sister Frances ("Fanny") to America following the death of their father, Patrick. It was the time of the great Potato Famine in the Emerald Isle, and the widow was hoping for a better life for the girls who had been raised in poverty.

The migrants first settled in Boston, but soon moved to Washington, DC. Nellie found a job as an elevator operator in a hotel that was frequented by government officials. It was said that one of the hotel guests was Ulysses S. Grant, who urged the comely Irish lass to go west. Whether following the future president's advice or their own initiative, the Cashmans moved to San Francisco.

Taking leave of her mother and sister, Nellie began a lifetime career of chasing the golden rainbow, following new strikes. She hired out as a cook in various Nevada mining camps, then opened a miner's boarding house in

Panaca Flat in 1872. It was the first of several camps where she opened businesses, always prospecting on the side, hoping to find her own pot of gold. In 1874 she joined the stampede to the Cassiar Mountains in British Columbia, establishing a boarding house at Telegraph Creek.

While there, she learned that twenty-six miners had become stranded in a remote location by a snowstorm. Distressed by reports that the army would not dispatch a rescue team, her Irish determination and her Catholic concern for others set her in motion. She hired six miners and lined up pack animals to take three-quarters of a ton of supplies to aid the stricken men. The army, afraid seven more people would be added to the casualty list, sent a squad of soldiers to stop and return the would-be rescuers. She defied the soldiers, refusing to go back.

After seventy-seven days of arduous travel through deep snow, the rescue party reached the snowbound miners. Instead of the twenty-six she expected, there turned out to be seventy-five. Most of them were suffering from scurvy. Nellie dosed them with Vitamin C and fed them fresh food, nursing them back to health and leading them to safety. When news of her sacrifice and success reached civilization, Nellie was dubbed the "Angel of Cassiar."

In 1879, Nellie returned to Arizona where there were newfound silver lodes. She opened Delmonico Restaurant then sold it the following year and moved to Tombstone. She first opened a store selling shoes and boots, the latter being made by her sister's husband. With that business going, she opened the Russ House Restaurant. Seeking publicity for the business, she contacted John Clum, publisher of the local newspaper, the *Epitaph*.

Clum, with whom she would reunite in Alaska two decades later, was highly impressed by the enterprising woman.

Fannie, whose husband had died and left her to raise their five children, joined Nellie in Tombstone that year.

Nellie's reputation as a compassionate and headstrong woman was heightened in 1884 after an outlaw quintet shot up the town, killing four people—one of them a pregnant woman and another a deputy sheriff. The outlaws were run down, jailed, tried in Tombstone, convicted, and sentenced to be hung. The case, known as the Bisbee Massacre, drew considerable interest. So many people were expected to view the execution that the sheriff had tickets printed up and passed them out in an attempt to limit the number of spectators. A businessman, seeing an opportunity for profit, built a set of bleachers for which he sold tickets.

Abhorring the morbid attitude of the populace and expressing her opinion that death should not be a cause for celebration, Nellie Cashman took meals to the convicted murderers—who professed to be Catholics and asked that their bodies be turned over to the priest. On the night before the hanging, someone led a group that tore down the bleachers. Nellie was strongly suspected of arranging the deed. When a fifth man was sentenced to prison for helping to plan the robbery that led to the murders, a mob broke the man out of jail and prepared to lynch him. Cashman drove her carriage into the angry crowd in an attempt to stop the lynching, but was unsuccessful.

Two years after the town's first legal hanging—and perhaps also the first illegal one—Nellie sold the Russ House. In that same year, Fannie died of tuberculosis;

Nellie took in the five orphaned children and raised them to become successful citizens. Her favorite nephew, Mike Cunningham, became a banker in Bisbee.

With the children tagging along, Nellie continued to follow rushes to boom camps in Nevada, Wyoming, Montana, and New Mexico. In each she opened a restaurant and prospected on the side.

The 1898 stampede could not be resisted by this woman, who by this time was fifty-five years old. She was not too old to climb the Chilkoot Pass, but was able to hire people to pack her goods up the golden stairs. True to form, she opened a restaurant and a mercantile store in Dawson, prospecting on the side.

By 1904, the boom in Dawson was subsiding. Miners were heading down the Yukon to Nome and the new settlement on the banks of the Chena, at Fairbanks. There, Nellie opened a grocery store. Three years later, Nellie went to the Koyukuk District on the eastern slope of the Nulato Hills and north of the Yukon. It was there that she found gold and mined a claim on Nolan Creek with good results. That became her base and she continued to be an independent woman. She became an accomplished dog musher and made headlines in 1922 when she mushed from Nolan, forty miles above the Arctic Circle, to Anchorage in the Southcentral part of Alaska. In 1923 the Associated Press featured her seventeen-day, three hundred fifty-mile trek from Nolan to Nenana. She was seventy-eight at the time.

Her health failing, she moved to Victoria, British Columbia, and was admitted to St. Vincent's Hospital where she died on January 4, 1925 at the age of eighty. It was that hospital she had helped build by raising money in 1874 and making contributions over the years thereafter. She had also played a major part in raising funds to build St. Joseph's Hospital in Fairbanks and helped build churches in Tombstone and elsewhere. It was in Tombstone where she convinced Wyatt Earp to open his Crystal Palace Saloon on Sundays so that church services could be held there until a church was built.

Nellie Cashman was called an Angel in many of the places where she was a part of the community. She was known for taking care of down-and-out miners, feeding and housing them in her establishments. She never married, but had an endless array of admirers among the many she helped.

Nellie Cashman spent lifetime in north, earning title "Angel" many times over.

Photo courtesy Library of Congress

'Wise Mike' Stepovich stows away, begins dynasty

An eighteen-year-old Slav christened Marko Stijpovic stowed away on a ship bound for the United States and went on to mine extensively on Yukon creeks. He was to father Alaska's last territorial governor.

Born in 1874 in Rasan, Montenegro (later to become Yugoslavia), the young man arrived in Baltimore, Maryland, in 1892, then traveled across the country to Fresno, California. There for five years, he raised cattle and harvested figs that were dried and sold to Nevada miners. By then he had attempted to make his name more pronounce-able, calling himself Mike Stepovich.

Hearing news of the Klondike gold discovery, in 1898 he headed north to seek his fortune. Arriving in Dyea he hired Native packers to take his extensive stock of goods over the Chilkoot. Encountering newly-arrived con man "Soapy" Smith, whose gang was collecting tolls from Klondike stampeders, he bypassed their camp in order to avoid paying the crooks. That accomplishment earned for him the nickname "Wise Mike," an appellation that stuck for the rest of his life—apparently deservedly so.

Arriving at Lake Bennett too late to proceed by way of the rivers, he built a cabin and wintered in the small town that sprang up on the lake's shores. When the lake broke up in the spring, he built a raft and drifted his supplies downstream. Encountering the rapids, he decided to load the freight onto a steamer to complete the trip to Dawson. That decision proved to be less than wise because the steamer was smashed in the rapids and the would-be outfitter's goods were lost. Undaunted, he acquired three draft horses and freighted others' goods between the camps during the winter of 1898-99. The following year he shifted operations to the Stewart River, prospecting on promising creeks between hauling opportunities. Two of his horses died during this time.

In July of 1902 Stepovich loaded his goods on the lone surviving horse, boarded a steamer and went to Circle where some claims were still being worked successfully. Once there he got word of the discoveries in the new Fairbanks district and set out cross-country to that destination. He obtained a site on Garden Island and, figuring that the spot's name was an omen, raised produce and grain.

The lure of gold was greater than that of farming, however, and the burly Balkan prospected and staked claims in the Fairbanks district. His first claim was filed on July 30, 1903, for a spot near the mouth of Fairbanks Creek and its confluence with Fish Creek. Over time, he purchased additional claims. During this period, Stepovich realized that many claims that had been staked were legally abandoned because assessment work had not been performed as required. He went to the recorder and over-filed to add these to his collection.

That action did not sit well with some of the original claimants, who considered that they were "jumped." Three of them—Merton Marston, William James, and "Poppa" John Butrovich—decided to right the supposed wrong. The trio showed up at Stepovich's cabin on Fish Creek and when their protests were rejected, one of the miners fired his rifle in Stepovich's direction. The intended target responded by emptying his six-shooter. None of the bullets found a human target but the intruders quickly abandoned their mission and sought refuge in a location they felt would be safer.

The three men were later taken into custody and charged with assault. Trial was held in November of 1909. Marston and James were found guilty and given light jail sentences, Butrovich exonerated.

Stepovich continued to mine while developing commercial property in the young city of Fairbanks. In addition to gold, "Wise Mike" was interested in other minerals, particularly tungsten. Scheelite, the ore from which tungsten is extracted, was found in both placer and lode form in the area. Tungsten is valuable because when added to steel it makes the result stronger.

To mine his claims more efficiently, Stepovich ordered huge dredges and heavy equipment. They were employed with success for the next two decades. His operations were shut down in the early days of the World War when the government commandeered his equipment to build airfields and defense installations.

Stepovich had become wealthy by 1918 and he married Olga Barta, a Croatian woman living in Oregon. Life in the wilds of Alaska did not suit her, however, and they were divorced. She took their son, Michael Anthony, who was born at St. Joseph's Hospital in Fairbanks in 1919, back to Oregon.

Wise Mike Stepovich married again in 1928 after a visit to his homeland. Over the years he had written his mother in the Slavic language. Because she could not read or write, the letters were read to her by Vuka Radovic, whose sister had married Stepovich's older brother. A romance blossomed and Vuka agreed to become the second Mrs. Stepovich.

Mike Stepovich the younger began writing his father and at age seventeen came to work with him. Young Mike went to school in Oregon and gained a law degree from the University of Notre Dame and after serving three years as a judge adjutant general in the United States Navy returned to Alaska where he practiced law in Fairbanks. He was elected to the Alaska Territorial House of Representatives in 1943, then served two terms in the Senate. Although he did not apply to fill out the term of Governor B. Frank Heintzleman, who resigned, Stepovich was recommended for appointment following a meeting with Secretary of Interior Fred Seaton. President Dwight D. Eisenhower appointed Stepovich, who was sworn in as the first Alaska-born governor on June 5, 1957. A staunch Statehood advocate, Stepovich was replaced in 1959 by William A. Egan, the first governor of the State of Alaska.

Wise Mike Stepovich died of a stroke in 1944 while living in California. He was sixty-five.

Nick and Evinda Tweet form lasting team of miners

An Alaska mining family operating on the Seward Peninsula today can trace its history back to its beginnings more than a century ago.

N. B. Tweet and Sons owes its founding to Nick Tweet who ventured over the Chilkoot Pass and rafted his way to the Fortymile District in 1894. Marrying a fellow Minnesotan five years later, the couple were to raise four sons and spend the rest of their lives making their living from material extracted from the frozen ground.

Nicholas Tweet was born August 14, 1871, in Franklin, Minnesota, the son of immigrants from Norway. His bride-to-be, Evinda Lyden, was born eight years later, on May 16, 1879, in Sacred Heart, Minnesota. She also was the child of parents from Norway. Both their birth towns are located in Renville County, but whether the two knew each other when growing up is unknown.

Tweet was not experienced in mining before he reached the Yukon, but he was a fast learner. His years at Fortymile were spent working for other miners and partnering with other prospectors as he searched for new ground that was promising.

He had not found good prospects before the fall of 1896 when word reached him of George Carmack's find on Rabbit Creek, upstream from his Yukon base. He wasted no time before joining the rush, but like so many others found all the creeks near the discovery had been staked. He worked for the successful stakers until 1898 when word of the Anvil Creek strike at the other end of the Yukon reached him.

Taking passage on a sternwheel steamer to St. Michael, he found that the discovery was across Norton Sound. He walked all the way around the inlet, arriving in Anvil City on June 30, 1899. He saw many tents packed along the shore, one shanty housing a saloon, and one log cabin. He managed to find showings of gold on the beach and did well enough to pay his way back to Minnesota that fall. There he persuaded Evinda to become his bride.

Leaving Evinda at Tacoma, Washington, Nick once more headed north. His wife remained in Tacoma and gave birth to two sons, Edgar and Norman, in 1900 and 1901, respectively. During those summers, Tweet found the Nome beach crowded with miners. Turning his attention farther north, he built a small boat and rowed it to Teller. He explored the Kougarok and Fairhaven districts, finding good prospects.

In 1902, Tweet took his wife and two sons to Teller. He, his wife and sons were to mine in the Port Clarence district until 1947. During that time two more sons, Harold and William, were born and when old enough joined in the work. When World War I broke out, graphite became valuable to the war effort and the Tweets developed an area where that substance was found in abundance. They shipped out five hundred tons of the material.

After the war ended, graphite was no longer of high value and the Tweets returned their attention to gold.

Edgar, the first-born son, married the Teller school teacher in 1927. He soon opened a trading post and in the Thirties sold his interest in the mines to other family members and expanded his store. Also in that decade, N. B. Tweet and Sons operated a dredge on Deese Creek until 1938.

The outbreak of World War II caused President Franklin D. Roosevelt to issue an executive order, L-208, shutting down all mining but that which was necessary to the war effort. Teller, the headquarters of the Tweet operations, was on the Bering Sea coast and was considered to be vulnerable to attack. The Alaska Guard was created, with Natives and other residents recruited to watch for potential invaders. Although aged seventy-one, Nick enlisted in the Alaska Territorial Guard as a private. The two older brothers also enlisted. Edgar, aged forty-two, was commissioned as a lieutenant and Norman, forty, was made a sergeant. Harold and Bill both enlisted and served in Alaska, Bill stationed in the Aleutians.

After the war, the Tweets turned their attention back to mining. Inasmuch as the Port Clarence claims were worked out, their headquarters was moved from Teller to Taylor, close to the center of the Seward Peninsula. The family was working on Humbolt Creek, continuing to mine successfully. Until 1970 they spent their winters in Teller. As recently as 2010, they launched a newly rebuilt and modernized bucketline dredge at Taylor.

Nicholas Tweet died July 14, 1967, followed by Evinda on May 28, 1975. Both died at their home at Taylor. Their sons now have also passed on and N. B. Tweet and Sons continues to carry on the family tradition under the guidance of Nick and Evinda's grandchildren.

Klondikers leave for Alaska, mainstays of 49th State

Many of those who made their way to the Klondike went on downriver to Alaska to seek their fortunes in the new strikes being found. Of those, quite a few remained in the land where they and their descendants became prominent citizens.

Two such were a Norwegian who turned to home improvement when he found no claims available and a Swede who worked his way over the Chilkoot Trail and founded a mining dynasty in Alaska. Enterprises begun by both are still going strong more than a century later.

Andrew Nerland, who was born March 18, 1870, in Molda, Norway, joined the stampede with two Anderson brothers, Herman and Louis. All had been in the painting and wallpapering trade and sought to get rich in the goldfields. They climbed the "Golden Stairs" in April of 1898, shortly after the Palm Sunday avalanche took the lives of some sixty-five or more adventurers.

Arriving too late to stake valuable claims for themselves, they turned to vocations with which they were familiar. The fledgling town of Dawson was bustling, buildings going up right and left. They found that occupants wanted to cover bare walls to make them more attractive. They quickly made arrangements to get supplies and hired out to make things cleaner and brighter. Nerland, in fact, was to make three arduous trips within a six-month period to obtain and bring back new stocks of paint, wallpaper and supplies over the pass.

Nerland soon brought his wife, Annie, to Dawson and there was born their son, Leslie.

By 1904 the boom was slowing in Dawson as big companies moved in to take over the claims and re-work them with heavy dredges. The partners decided to move downriver where a new strike had resulted in what apparently was to become a permanent settlement at Barnette's Cache on the Chena River.

Nerland sent his wife and son to Seattle while he and his partners became established. Anderson Brothers and Nerland opened in the new town of Fairbanks, selling wallpaper, paint, signs, glass, and doors and sashes. They also contracted out as painters and paper-hangers. Before long, they opened new stores in Nenana and Iditarod. Their Fairbanks store had living quarters upstairs and Annie and Leslie came to live there.

Nerland became a major figure in Fairbanks and held leadership positions in the Masonic Lodge, Odd Fellows, Pioneers of Alaska, and the Presbyterian Church. He was elected to the city council, as mayor, and seven times was elected to seats in the territorial Legislature.

Nerland was among those who helped bring to Alaska an institution of higher learning. The Alaska Agricultural College and School of Mines was established in 1917 and became the University of Alaska in 1935. Nerland was a member of the Board of Regents and once signed for a ten thousand-dollar loan to help keep the institution going during rough times.

In 1955, Les Nerland was chosen as a member of

Alaska's Constitutional Convention, convening at the University of Alaska to prepare a constitution that would govern the northern territory should statehood be granted. The constitution was ratified by voters in 1956. It became official two years later when President Dwight D. Eisenhower signed the Alaska Statehood Bill and the Forty-ninth star was added to the Stars and Stripes.

According to reports, on February 6, 1956, Les showed the newly-drawn document to his father, eighty-five, and who many years earlier submitted a legislative measure seeking statehood for Alaska. "It's a big step toward statehood," the elder Nerland said. Fifteen minutes later, he collapsed and died of a heart attack.

Until late in the Twentieth Century, the Nerland family operated furniture stores throughout Alaska and in Seattle. They since reduced the number of stores and scope of the business. Members of the sixth generation of Nerlands are shown today in commercials for the stores.

Born September 27, 1875, **David Strandberg** came to America twenty years later. He worked in coal mines in Pennsylvania and Minnesota before moving to Shelton, Washington. He was engaged in logging and attended college in Seattle.

In 1898, Strandberg and his brother Charley went to Skagway, en route to the Klondike. To help with the cost of meeting Canadian requirements for a year's worth of supplies for each person, they hired out to help pack gear over the Chilkoot Pass for others who joined the Gold Rush.

In Dawson, the brothers partnered with Alex Matheson and Jim Lund to mine Number Five Below on Dominion Creek. While that creek did not produce the kind of wealth that some others did, the partners managed to live comfortably. While in Dawson they met Girard Johanson, his wife and nieces Jenny and Hilma.

As had Nerland and so many others, they headed for Fairbanks in 1905 after their own mine was worked out. Three years later, David Strandberg and Jenny Johansen were married. They mined on Ester Creek, outside Fairbanks.

From Fairbanks, Strandberg followed reports of a new strike on Otter Creek in what was to become the city of Flat, near another boom town of Iditarod.

In 1932 David's four sons joined him in forming Strandberg and Sons Mining Company, headquartered in Anchorage. They operated gold mines on Cripple, Candle, and Utopia creeks and also a platinum mine in the Goodnews Bay area.

David Strandberg died in 1951, Jenny in 1954. Both are buried in the Anchorage Memorial Cemetery.

The Nerland and Strandberg families are but two of many whose roots go back to Gold Rush days. Their stories would add many pages to Alaska's history.

Reporter Marston makes million in mining

When Merton H. Marston, city editor of the *Daily Herald* in LaPorte, Indiana, heard that some men from that city were heading for the Klondike, he decided to throw in with them. With him in the syndicate were James W. Davis, a civil engineer; John Merton Davis, a blacksmith; and Eugene G. Neff, a carpenter. They were joined while en route by George McDonald, whose occupation is unknown.

In 1898 when they set out, Marston was a twenty-seven-year-old.

The syndicate reached Dawson, but it is unclear where they mined and whether they stayed together.

By 1906, however, Marston was known to be on Fairbanks Creek, mining and operating a saloon. Two years later, he had returned to his earlier profession and was writing for the *Weekly Fairbanks News*, which was to become the *Fairbanks Daily News-Miner,* today Alaska's oldest daily newspaper.

In Fairbanks, Marston met John Beaton and William A. Dikeman. The two had prospected on Otter Creek, in what would be known as Flat. Marston grub-staked them for one-sixth share in what their mine was to produce.

When it came time to settle up, Marston's share was

worth almost one million, seven hundred thousand dollars. Marston quit his newspaper job and left Alaska for Seattle where he and his wife, Stella, built a hotel. Later they moved to Oregon and built another hotel, which burned. They relocated to another Oregon city and opened the Rendezvous Restaurant. Marston died in April of 1935.

John T. Erickson mined, made more with livestock

John T. Erickson left his young family to join the Yukon Gold Rush early in the 1900s. He made a small fortune mining, bought a herd of horses and mules, returned to Nome and made even more. With his gains, he settled in Pueblo, Colorado, and in 1906 built the Erickson Block at a cost of forty thousand dollars.

The building was a showcase, featuring marble imported from Italy. Italian sculptors were brought over to carve the material into decorative fixtures. Even though the three-story building had a bank as the anchor tenant, the cost of the ornate structure could not be recovered and its owner lost his investment.

Erickson was born in 1865 in Chicago, Illinois, the son of Swedish immigrants Andrew John Erickson and Marie Katherine (Parson) Erickson. He married Ida Anderson on December 14, 1885, in Denver, Colorado. They had five children, Alma, Edith, Ernest, Alexander, and Walter. Ida Erickson died in 1898 at age thirty-two and John married Elna A. Bong on April 18, 1908, in Denver, Colorado. They had five children, Mildred, Richard H., John Andrew, Lloyd, and Vernon.

Walter, whose daughter Barbara is married to the author, followed his father's footsteps and mined in the

Nome area in the 1950s. An inventor, he patented a "hinged parallelogram" to be the undercarriage of a dryland dredge. He formed a corporation which leased several claims in the Nome area, with exploration work done in 1953. A planned operation to recover Scheelite (tungsten) and gold was thwarted when the government cancelled a contract to purchase the strategic mineral.

There is no known record of John T. Erickson's activities in Alaska. A hotel register lists him and gives his occupation as "hauler." Newspapers from the Gold Rush that have been reviewed make no mention of him.

John Erickson prospected throughout the West and settled in Pueblo. He operated a rock quarry and also acquired a sawmill. Prior to his Alaska venture he farmed near Courtland, Colorado. He discovered a thick vein of coal and quietly set about preparing to mine. In short order he had a large number of men working the claim. A settlement grew up around the mine and was called Coal City. The operation was halted a few years later when drillers struck salt water that flooded the mine and spilled into a local stream with serious environmental results.

Following the mine disaster Erickson prospected the nearby streams, leaving his farm and children in the care of his brother-in-law, Charley Sallman.

A 1976 article in the Scandia *Journal* described Erickson as a "promoter." It was a term his son Walter later used with disdain when describing get-rich schemers and high-pressure sales people. He advised his children not to dig into his father's past as it might disclose undesirable facts. Walter was only three when his mother died and he and his siblings were handed over to relatives to be raised. The elder Erickson apparently left for

Nome soon after his first wife died. An unusual coincidence is that Ida and both of her daughters all died at the age of thirty-two. The older daughter, Alma, an Army nurse, died of influenza. Cause of death of the others is not known.

In 1908, a decade after Ida died, John Erickson married Elna A. Bong. They were to have five children, all of whom lived with their mother in Wisconsin. He died in Monte Vista, Colorado, but the year of his death and place of burial are not known.

Using $40,000 in earnings from the Nome Gold Rush, John T. Erickson built this stone structure at the corner of Northern Avenue and East Evans Avenue in Pueblo, Colorado. He was unable to recover the construction cost from rental income, and ended up losing his investment.

Photo courtesy Pueblo Public Library

Bibliography

Hundreds of books have been written about the Yukon River and the gold rushes to both ends of that two-thousand-mile stream. For a deeper look at the history of that area and the people who colored that story, the author recommends the following from among his sources for this book.

Alaska, a History of the 49th State, by Claus-M. Naske and Herman E. Slotnick, University of Oklahoma Press.

The Sourdough and The Queen, the Many Lives of Klondike Joe Boyle, by Leonard W. Taylor, Merheun Publications, Agincourt, Ontario, Canada.

The Klondike Nugget, by Russell A. Bankson, The Caxton Printers, Caldwell, Idaho.

Klondike Fever, by Pierre Berton, Basic Books, New York.

The Klondike Stampede, by Tappan Adney, Harper, New York.

Alaska, a Bicentennial History, by William R. Hunt, W. W. Norton & Company, Inc., New York.

I Married the Klondike, by Laura Beatrice Berton, Harbour Publishing, Madeira Park, British Columbia, Canada.

Trail to North Star Gold, A True Story of the Alaska-Klondike Gold Rush, by Ella Lung Martinsen, Ninsford & Mort, Publishers, Portland, Oregon.

The Gold Fields of the Klondike, Fortune Seekers Guide to the Yukon Region of Alaska and British America, by John W. Leonard, A. N. Marquis & Company, Chicago.

Crooked Past, the History of a Frontier Mining Camp: Fairbanks, Alaska, by Terrence Cole, University of Alaska Press, Fairbanks.

Diary of an Alaskan Madam, by Lorina Ewing, Lorina's Touch Publishing, Durango, Colorado.

Wyatt Earp and Coeur d'Alene Gold!, Stampede to Idaho Territory, by Jerry Dolph and Arthur Randall, Museum of North Idaho.

'Stroller' White, Tales of a Klondike Newsman, compiled by R.N. DeArmond, Mitchell Press, Ltd., Vancouver, B.C.

Gold Rush Nome, by Carrie M. McLain, Graphic Arts Center, Portland, Oregon.

Good Time Girls of the Alaska-Yukon Gold Rush, by Lael Morgan, Epicenter Press, Fairbanks and Seattle.

History of Alaska 1730-1885, by Hubert Howe Bancroft, Hafner Publishing Company, Darien, Connecticut.

Pioneering the Yukon, 1892-1917, by Anna DeGraf, Archon Books, Hamden, Connecticut.

Gold Rush Women, by Claire Rudolf Murphy and Jane G. Haigh, Alaska Northwest Books, Anchorage.

The Klondike Letters, by Jack Beddall

The Last Grand Adventure, by William Bronson with Richard Reinhardt, McGraw-Hill Book Co., New York

A Klondike Scrapbook, Ordinary People, Extraordinary Times, by Norman Bolotin, Chronicle Books, San Francisco.

Klondike Lost, a Decade of Photographs by Kinsey & Kinsey, Norman Bolotin, Alaska Northwest Publishing Company, Anchorage.

Building Alaska with the United States Army, 1867-1965, U.S. Army Alaska Pamphlet 360-5

Report of the U.S. Revenue Cutter Bear and the Overland Expedition, by the U.S. Revenue Cutter Service, U.S. Government Printing Office, 1899

A Klondike Scrapbook, Ordinary People, Extraordinary Times, Norman Bolotin, Chronicle Books, San Francisco.

Klondike Lost, a Decade of Photographs by Kinsey & Kinsey, Norman Bolotin, Alaska Northwest Publishing Company, Anchorage.

Old Yukon Tales, Trails, and Trials, by James Wickersham, University of Alaska Press, Fairbanks

Klondike Tale, Jack London, The Modern Library, New York

Jack London's Stories of the North, Di Lernin Publishers

Newspaper archives:

The Dawson News

The Skagway News

The Klondike Nugget

Nome Gold Digger

The Nome Nugget

On-Line Sources:

Alaska Mining Hall of Fame

The Postal Museum

University of Washington Libraries

National Park Service

Canadian Parks

Canada Yukon Archives

Index

Model of washing plant designed by Walter Erickson using his patented "hinged parallelogram" undercarriage. Intended to be erected on or near bedrock, material to be processed using Dorrco jigs is fed by a dragline into the hopper where it would be thoroughly washed and screened. The plant would ride on rails, moving laterally to follow the width of the paystreak. With the weight concentrated at one end, the opposite skid would be advanced by a bulldozer. Tailings would be discharged and distributed behind the operation. A dedicated placer miner, Erickson explained that recovery of fine values would be increased compared to a floating dredge due to material being handled "dry" rather than lifted through water. Another advantage would be that tailings would be spread by movement of the plant, thus avoiding separate handling. He planned to utilize the plant at Nome in the 1950s to recover tungsten and gold but had to drop those plans when the government cancelled its contract to buy the tungsten, a strategic mineral.

Photo by Bob Boyer from Lee Jordan archives